Cambridge Elements

Elements in Organizational Response to Climate Change
edited by
Aseem Prakash
University of Washington
Jennifer Hadden
University of Maryland
David Konisky
Indiana University
Matthew Potoski
UC Santa Barbara

CORPORATE CLIMATE ADAPTATION

Translating Complex Societal Risks into Business as Usual

Maria-Therese Gustafsson
Stockholm University
Lisa Dellmuth
Stockholm University

Shaftesbury Road, Cambridge CB2 8EA, United Kingdom

One Liberty Plaza, 20th Floor, New York, NY 10006, USA

477 Williamstown Road, Port Melbourne, VIC 3207, Australia

314–321, 3rd Floor, Plot 3, Splendor Forum, Jasola District Centre,
New Delhi – 110025, India

Cambridge University Press is part of Cambridge University Press & Assessment,
a department of the University of Cambridge.

We share the University's mission to contribute to society through the pursuit of
education, learning and research at the highest international levels of excellence.

www.cambridge.org
Information on this title: www.cambridge.org/9781009780605
DOI: 10.1017/9781009780636

© Maria-Therese Gustafsson and Lisa Dellmuth 2026

This publication is in copyright. Subject to statutory exception and to the provisions of relevant collective licensing agreements, with the exception of the Creative Commons version the link for which is provided below, no reproduction of any part may take place without the written permission of Cambridge University Press & Assessment.

An online version of this work is published at doi.org/10.1017/9781009780636 under a Creative Commons Open Access license CC-BY-ND 4.0 which permits re-use, distribution and reproduction in any medium providing appropriate credit to the original work is given. You may not distribute derivative works without permission. To view a copy of this license, visit https://creativecommons.org/licenses/by-nd/4.0

When citing this work, please include a reference to the DOI 10.1017/9781009780636

First published 2026

A catalogue record for this publication is available from the British Library

A Cataloging-in-Publication data record for this Element is available from the Library of Congress

ISBN 978-1-009-78060-5 Hardback
ISBN 978-1-009-78058-2 Paperback
ISSN 2753-9342 (online)
ISSN 2753-9334 (print)

Additional resources for this publication at http://www.Cambridge.org/Gustafsson

Cambridge University Press & Assessment has no responsibility for the persistence or accuracy of URLs for external or third-party internet websites referred to in this publication and does not guarantee that any content on such websites is, or will remain, accurate or appropriate.

For EU product safety concerns, contact us at Calle de José Abascal, 56, 1°, 28003 Madrid, Spain, or email eugpsr@cambridge.org

Corporate Climate Adaptation

Translating Complex Societal Risks into Business as Usual

Elements in Organizational Response to Climate Change

DOI: 10.1017/9781009780636
First published online: February 2026

Maria-Therese Gustafsson
Stockholm University

Lisa Dellmuth
Stockholm University

Author for correspondence: Maria-Therese Gustafsson,
Mariatherese.gustafsson@statsvet.su.se

Abstract: As private companies assume a growing role in climate adaptation, their strategies may harm society and ecosystems unless grounded in responsible business conduct. This Element offers a new perspective on responsible business conduct in climate adaptation, presenting a theoretical framework that explains how regulatory and political factors external to firms influence their consideration of societal needs when adapting to climate change. Using a novel quantitative and qualitative dataset, the Element shows that the world's largest mining companies have primarily addressed climate risks through conventional corporate social responsibility strategies rather than procedural components of responsible business conduct, such as risk assessments, participation, and transparency. The results suggest this outcome is best explained by a combination of weak governance, lax voluntary standards, and civil society advocacy. This title is also available as Open Access on Cambridge Core.

Keywords: climate adaptation governance, extractive industries, transnational companies, corporate accountability, responsible business conduct

© Maria-Therese Gustafsson and Lisa Dellmuth 2026

ISBNs: 9781009780605 (HB), 9781009780582 (PB), 9781009780636 (OC)
ISSNs: 2753-9342 (online), 2753-9334 (print)

Contents

1 Introduction: Why Corporate Adaptation Matters   1

2 A Theoretical Framework for Studying Responsible Business Conduct in Adaptation   14

3 Mapping Responsible Business Conduct in Adaptation   24

4 Explaining Irresponsible Business Conduct in Adaptation   36

5 Conclusion: The Future of Responsible Corporate Adaptation   59

References   68

Online Appendix available at http://www.Cambridge.org/Gustafsson

1 Introduction: Why Corporate Adaptation Matters

Heat waves, wildfires, water scarcity, and extreme rainfall illustrate the wide-ranging and often devastating impacts of climate change on both human societies and ecosystems (Begum et al. 2022: 144). Adapting to these risks is increasingly recognized as one of the most pressing problems of our time. According to the World Economic Forum's 2022 annual survey of business and government leaders, climate action failure is the greatest anticipated threat companies face in the coming decade (World Economic Forum 2022). Companies and policymakers alike tend to frame climate adaptation narrowly as a physical risk to individual firms that could reduce productivity and disrupt supply chains. As a result companies typically address adaptation as an internal matter of business resilience, rather than as a societal risk (TCFD 2017). In contrast, the failure of companies to decarbonize at the scale and speed required to avoid socio-environmental harm has been framed as a critical societal concern in both scholarly and policymaking debates.

In this Element, we challenge the conventional view that corporate adaptation is solely, or primarily, a matter of business resilience, as corporate adaptation (or lack thereof) has far-reaching consequences for nature and society. Failure by companies to effectively adapt to climate change can have catastrophic socio-environmental consequences. Consider the example of the Arctic oil spill caused by the Russian company Nornickel in 2020 – recognized as the worst Arctic oil spill in history (The Moscow Times 2020). The company had not implemented any adaptation measures despite well-documented climate risks associated with melting permafrost. In the words of a Greenpeace representative working in the Arctic region, Nornickel's responsibility for its impacts on society and the environment are clear: "It is a powerful company [...] We have known about the melting permafrost for at least 10 years. There should be monitoring systems, there should be special constructions that takes climate risks into consideration in the work it carries out in areas like this" (Interview 33).

This Element defines corporate adaptation as "the process of adjustment by companies to actual or expected climate change and their associated effects through changes in business strategies, operations, practices, and/or investment decisions" (cf. IPCC 2022: 5). Companies have often relied on "technical fixes," such as air-condition installations, back-up power installations, early warning systems, and water infrastructure to adapt to climate change (Goldstein et al. 2019; cf. Nightingale et al. 2020). While these may appear to be simple technological adjustment aimed at climate-proofing facilities, they are often highly complex and may carry unforeseen societal consequences. Even when companies engage in climate adaptation, their actions can fall short of addressing the full magnitude of these risks and may not reflect or even conflict with societal priorities and needs (Grabs et al. 2026).

For example, wine producers have started relocating vineyards to higher elevations to maintain their production, which in turn has reduced freshwater availability (Hannah et al. 2013). In Brazil, large-scale agricultural producers have implemented irrigation schemes to cope with climate-induced water stress that have led to conflicts with local communities over water access and food security (Gustafsson et al. 2024). In Chile, mining companies and local communities have suffered from unprecedented megadroughts, leading companies to invest billions of dollars in desalination plants using seawater. While these plants have helped reduce hydrological insecurity and mitigate social conflicts at mining sites, they have also created new environmental problems and social unrest in coastal areas (Odell 2021). These examples underscore how corporate adaptation (or the absence thereof) can exacerbate vulnerabilities and produce socio-environmental harms.

This Element asks how companies address the societal risks associated with their business activities in a context of climate change and what motivates them to do so. We develop a theoretical framework that, firstly, conceptualizes responsible business conduct (RBC) in the context of adaptation and, secondly, identifies central regulatory and political factors that shape RBC in adaptation. We explore the framework in the mining sector, which is especially vulnerable to climate impacts, and where RBC is crucial for preventing and mitigating the negative socio-environmental consequences of multinational corporate activity.

1.1 A Framework for Analyzing Corporate Responsibility in Adaptation

Our argument begins with the premise that companies can reduce climate-related harms associated with their business activities by integrating principles of RBC. In line with the OECD's Guidelines for Multinational Enterprises on Responsible Business Conduct (OECD 2023), we refer to RBC as the behavior of businesses to avoid and address adverse impacts of their operations, while contributing to sustainable development in the countries where they operate. In this Element, we focus on four RBC principles central to climate adaptation: assessments of climate risks to society, stakeholder participation, transparency, and corporate social responsibility (CSR) investments in public adaptation goods. In doing so, we build and advance on insights from CSR, Business and Human Rights, and climate adaptation governance scholarship.

Despite growing attention to RBC in relation to environmental matters in general, climate adaptation continues to be only marginally addressed in emerging voluntary and mandatory frameworks. For instance, the voluntary OECD guidelines (2023: 37) stipulate that "enterprises should avoid activities, which undermine climate adaptation for, and resilience of, communities, workers and ecosystems," leaving it unclear what this means in practice. Recent Human Rights and Environmental Due Diligence (HREDD) legislations, such as the Corporate Sustainability Due Diligence Directive (CSDDD) and the German Supply Chain Due Diligence Act (LkSG) law, make RBC obligations legally binding. However, these legislations do not cover climate adaptation, meaning this conduct is currently not mandated by law.

Further, scholars have not yet systematically examined RBC in the area of adaptation. We suspect that there are two reasons for this. First, there is a lack of a common understanding of what adaptation means for corporations (Averchenkova et al. 2016), and the concept is often used to refer to a variety of measures to address climate-related risks. Second, we argue that, compared to other RBC issues, corporate responsibilities vis-á-vis society are particularly ambiguous when it comes to adaptation. Climate-related risks can intersect with and exacerbate the impacts of business activities (Odell et al. 2018). In such situations of "double exposure" to both climate impacts and economic activities, the potential socio-environmental consequences are not only likely to be particularly severe (Leichenko and O'Brien 2008), but it is also more difficult to define the boundaries of corporate responsibilities in addressing these intersecting impacts. As climate-related impacts become more frequent and severe, we argue that companies must be better prepared and held accountable for how they respond to these intersecting impacts.

This Element seeks to remedy this shortcoming in the existing literature. The remainder of this section lays out the argument and introduces the research design. In the following two subsections, we briefly summarize our theoretical framework, which will be discussed in greater detail in Section 2.

Conceptualizing Corporate Responsibility in the Context of Climate Adaptation

This Element identifies central procedural and distributive RBC principles in the context of adaptation. These principles can help to prevent, or at least mitigate, the adverse effects of business activities amid climate change. Clarifying these responsibilities is important, as it facilitates more transparent and inclusive corporate responses to climate risks and supports the development of robust public policies and accountability mechanisms that align corporate adaptation strategies with societal needs. We begin by discussing the three procedural principles, followed by the distributive principle.

Assessments of climate risks to society enable companies to identify, prevent, and develop measures to address potential climate-related harms associated with their business activities before they occur, rather than merely reacting after the damage has already been done. Such climate risk assessments for society refer to an adequate evaluation of the societal risks associated with companies' activities in a context of climate change, which may lead to restrictions of business activities. While many companies already assess their own exposure to climate risks, RBC requires a shift in focus toward the socio-environmental risks stemming from their business activities and adaptation strategies. Companies can achieve this by integrating climate adaptation into existing risk management processes, such as Environmental Impact Assessments (EIAs) and HREDD. More generally, human rights and environmental risk assessments are a core component of soft and hard law frameworks on RBC, such as the United Nations Guidelines for Business and Human Rights (UNGPs) (OHCHR 2011), the OECD Guidelines (2023), as well as binding laws building on these frameworks.

Through meaningful stakeholder *participation*, affected groups can influence how their perspectives and adaptation needs are integrated into corporate adaptation responses. By participation we mean the engagement of relevant, affected stakeholders at various stages of adaptation interventions (Eriksen et al. 2021), including the assessments of climate risks, the planning of responses, and the implementation of adaptation interventions. In the context of the extractive industries, companies can, for example, integrate climate adaptation into EIAs and participatory water monitoring, thereby leveraging their existing institutionalized processes for stakeholder involvement. Stakeholder involvement throughout the process of assessing and addressing environmental and human rights risks is a cornerstone of the soft law frameworks of RBC, such as the UNGPs and the OECD Guidelines.

Transparency is important, as it can empower affected groups and governmental agencies to critically evaluate the risks associated with business activities in a context of climate change. This is particularly important in high-risk situations, where such activities may expose surrounding communities and environments to severe risks. We define transparency as companies' obligation to inform affected groups and governmental agencies about their exposure to climate risks and the adaptive measures they have taken to address them (cf. OECD 2023). Along with risk assessment and stakeholder participation, transparency about risks associated with business activities and how companies monitor and respond to them is another core component of the soft law frameworks of RBC.

Finally, *CSR investments* can enable companies to move beyond a "do no harm" approach and make positive contributions to adaptation in the countries where they operate. This principle falls into the category of resource distribution, rather than the procedurally oriented principles outlined in HREDD frameworks. CSR investments typically involve the provision of public adaptation goods such as early warning systems, capacity building, and water infrastructure. When channeled through partnerships with public authorities and civil society organizations (CSOs), CSR can help build institutional capacity and support the implementation of adaptation plans (cf. UNEP 2021). However, companies often use CSR investments as strategic tools to legitimize their business activities and buy off opponents. In such cases, CSR can lead to the fragmentation of CSOs and the closure of public debates on climate-related risks and adaptation solutions that companies perceive as threatening to their interests (cf. Amengual 2018; Gustafsson 2018).

In sum, if meaningfully integrated into corporate adaptation responses, these four RBC principles promise preventing, or at least mitigating, the adverse impacts of corporate adaptation, and clarifying where restrictions of business activities are needed. As we elaborate in Section 2, our framework leaves out RBC principles in international frameworks, such as corporate governance and access to remedy, which are important in general but which is beyond the scope of our inquiry. Instead, we focus on those principles that are particularly relevant for preventing harm and developing regulatory policies addressing the intersecting impacts of climate change and business activities. Companies are unlikely to integrate these principles and change their behavior unless there are strong external incentives and/or regulatory pressures in place.

Regulatory and Political Factors Shaping Responsible Business Conduct in Adaptation

The second part of our theoretical framework explains how regulatory and political factors shape RBC in adaptation. We privilege three factors: public governance, civil society advocacy, and participation in voluntary adaptation initiatives. The private governance literature has highlighted that these factors matter for companies' more general sustainability practices (Bartley 2018b; Lambin and Thorlakson 2018; Thorlakson et al. 2018a, 2018b), but to date they have not been examined in relation to climate adaptation. We build on this debate by theorizing how these external factors, both individually and in combination, affect how companies incorporate RBC into their responses to climate risks.

Public governance can influence RBC through soft steering and legal requirements. We refer to public governance as any hard or soft regulatory activities aimed at fostering RBC. For example, governments can introduce legal requirements to assess and address climate risks in different types of sectoral policy instruments, such as EIAs and water licenses. In some cases, this may lead to restrictions for business activities linked to substantial socio-environmental risks (cf. Bartley 2018a; Odell et al. 2018). Governments tend to be reluctant to enforce stringent regulatory controls on businesses and instead use partnerships, support, and soft steering to influence company behavior (cf. Abbott and Snidal 2009; Eberlein 2019). To effectively steer and regulate company behavior, public authorities require capacity and robust monitoring systems to be able to assess the impacts of business activities in the context of climate change (cf. Lambin and Thorlakson 2018).

In the absence of robust governance frameworks, *voluntary adaptation standards* can contribute to diffusing norms and contribute to learning about RBC in adaptation (cf. Grabs 2020). Such standards include multistakeholder initiatives, certification schemes, and private standards that set requirements for RBC (cf. Marx et al. 2024). Yet, a large body of research shows that these standards frequently lack stringency, are unevenly applied, and are poorly incorporated into firms' management structures (Bartley 2018b; LeBaron et al. 2017). Unlike climate mitigation, adaptation is covered by relatively few voluntary standards, and it remains unclear to what extent and how effectively these existing standards require companies to incorporate RBC principles into their adaptation responses.

Civil society advocacy can influence RBC by pressuring companies to develop more transparent and inclusionary adaptation responses. We refer to civil society advocacy as a civil society-based form of regulation of companies (cf. Newell 2008), in which different types of CSO actors rely on campaigns, partnerships, and public scrutiny to compel companies to address socio-environmental risks in their adaptation strategies. Civil society has played a key role in compelling companies to comply more meaningfully with human rights and environmental standards (Bartley 2018b; Thorlakson et al. 2018a). However, it has often proved challenging for civil society to gather solid evidence of corporate impacts in the area of adaptation (Gustafsson et al. 2024). Additionally, unlike the large-scale protests against fossil-fuel emissions, mobilization around climate adaptation remains relatively rare (de Moor 2022).

We do not claim that these are the only factors likely to influence RBC in adaptation. Existing corporate adaptation research has also highlighted the role of biophysical risk exposure (Averchenkova et al. 2016; Mbanyele and Muchenje 2022) and firm characteristics (Klein et al. 2022; Pinkse and Gasbarro 2019). In contrast, our theoretical framework is embedded in the private governance literature and privileges regulatory and political factors, which to date remain under-researched in the context of adaptation. Next, we lay out our research design to empirically examine our account.

1.2 Evidence from the Mining Sector

We examine our theoretical framework by using a new quantitative and qualitative dataset on corporate adaptation in the mining sector. This sector suits our inquiry for several reasons. First, the mining sector is extremely vulnerable to different types of environmental hazards and climate-related risks. Extreme weather events can cause floods and droughts that can damage mining infrastructure containing highly toxic waste and in turn lead to land and water contamination (Phillips 2016). Additionally, mining is one of the most water-intensive industries. Unless companies adequately adapt to climate risks like these by climate-proofing their infrastructure and reducing their water consumption, surrounding communities can be severely affected by water stress.

To illustrate the mining sector's vulnerability to climate-related risks, we created data to visualize the context in which corporate adaptation in the mining sector unfolds. Figure 1 is a map showing both mining dependency and climate vulnerability. "Mining dependency" applies to countries where metals and minerals account for more than 20 percent of exports by value and mineral rents constitute more than 10 percent of the country's gross domestic product (GDP). The data is drawn from the Mining Contribution Index compiled by the International Council on Mining and Metals (ICMM), a leading industry association. We measure climate vulnerability by using the Physical Vulnerability to Climate Change Index. This index captures the main

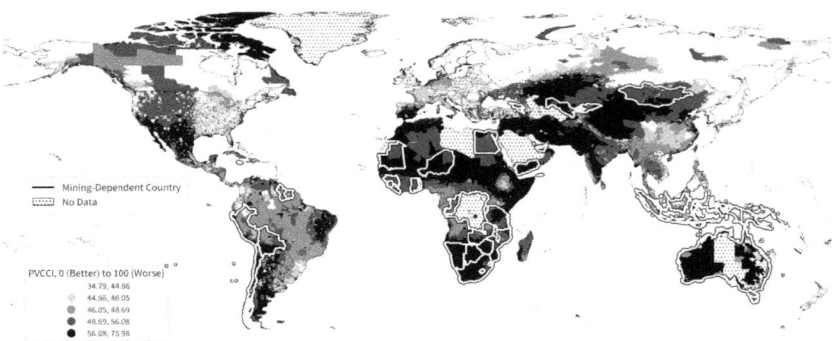

Figure 1 Climate vulnerability and mining dependency.
Source: ICMM (2021, 2022); Mining Contribution Index; FERDI (2022).

structural and physical consequences of climate change that are likely to affect both local communities and mining operations, underlining the centrality of RBC in adaptation in the affected areas. The climate index captures major risks for livelihoods posed by flooding, aridity, temperature shocks, rainfall shocks, and cyclones (Feindouno et al. 2020; FERDI 2022; Goujon et al. 2022).

The map also shows the overlap of mining dependency and climate vulnerability, delineated by the occurrence of mining dependence in the darker-shaded geographical areas. In several mining-dependent countries, such as Namibia, South Africa, and Yemen, the entire country is characterized by high levels of climate vulnerability. In other mining-dependent countries, such as Australia, Mongolia, and Peru, subnational areas are varyingly vulnerable to climate change. Taken together, this information clearly supports the argument that the mining industry is a high-risk sector where RBC in adaptation is critical to prevent and mitigate severe socio-environmental harm.

Moreover, the mining sector is an important case, as many companies have voluntarily adopted RBC policies to manage their relationships with host communities. While practices vary widely across companies, the most advanced companies usually have policies and procedures for socio-environmental risk assessment, information disclosure, participation, and often make CSR investments to showcase their positive contribution to local development. Thus, we expect companies to build upon and integrate climate adaptation into existing RBC policies and procedures developed to govern interactions with local communities and state agencies concerning the management of socio-environmental risks. At the same time, a substantial body of literature has found that RBC principles are often poorly implemented in the mining sector. For example, EIAs and participatory processes have frequently failed to provide local communities either with access to relevant information about socio-environmental risks or meaningful participation in decision-making processes involving large-scale mining projects (Murguía and Bastida 2024). Furthermore, companies have used CSR investments to buy off opponents, aiming to avoid costly conflicts with local communities (Amengual 2018; Franks et al. 2014).

Consequently, the mining sector represents a case where companies can build on existing RBC policies, but at the same time where there are significant barriers related to poor public governance and ineffective voluntary standards. By investigating how companies in the mining sector implement the RBC principles in adaptation, we provide important lessons with applicability to other sectors. To analyze RBC in adaptation, we developed a new qualitative and quantitative dataset.

Mapping Responsible Business Conduct in Adaptation through a Quantitative Dataset

To explore the first part of our theoretical framework, we descriptively map the extent to which companies have integrated the RBC principles into their adaptation responses. For this purpose, we compiled a dataset based on a qualitative coding of documents from the world's thirty-seven largest mining companies according to their market capitalization in 2019 (PWC 2019; see Appendix A; Table A1). An inductive coding of all available documents from 2017 to 2019 resulted in a company-level dataset with the capacity to descriptively map companies' perceptions of climate risks, adaptation responses, and RBC (see Appendix B; Tables B1, B2, and B3). The corporate documents we compiled for our research include annual reports, sustainability reports, reports to the Carbon Disclosure Project, Global Reporting Initiative reports, as well as policies and reports available on the companies' websites. The time period covered begins in 2017, prior to which the observed companies disclosed little or no information on their adaptation responses, conditions that rendered them ill-suited for analysis.

To operationalize the RBC principles, we confronted the challenge that many companies do not explicitly label their responses to climate risks as "adaptation." Rather, they tend to use several terms, like flood risk management, risk assessments, and resilience, somewhat interchangeably (see also Averchenkova et al. 2016). This makes it difficult – but not impossible – to identify different types of activities that would here be classified as "adaptation" (Dupuis and Biesbroek 2013). Thus, we examined the documents in conjunction with the interviews described later to operationalize the four principles (see also Section 3). More specifically, we coded whether companies adhered to the four RBC principles, as we describe in detail in Section 3 and Appendix B; Table B3. We also used the documents to measure a range of other aspects, including whether companies framed climate-related risks as economic and/or societal, and how they responded to these risks organizationally. This is useful information to set the stage for the mapping of the RBC principles.

Relying on company self-reporting of corporate adaptation has the advantage that we can generate this quantitative dataset, but as with any measurement approach, it also has limitations. Companies' reporting is often fragmented, biased, and vague and it should, therefore, be treated as a strategic communication tool employed by companies to present themselves as leaders in upholding voluntary sustainability standards. Therefore, the quantitative document analysis may overestimate RBC in adaptation and must be interpreted with caution.

To provide a more balanced analysis, we complement the quantitative data with eighty-seven semi-structured interviews with representatives of companies, business associations, civil society, and the public sector during the period 2020–2022. We transcribed and coded all interviews in Atlas.ti. We conducted forty-one of the eighty-seven interviews in relation to our dataset of the world's largest mining companies (see Appendix C). The main aim of these interviews was to contextualize the quantitative coding of the RBC principles. In this way, this Element offers a context-sensitive and critical analysis of how corporate actors perceive climate risks, adapt their practices, and adhere to RBC principles when adapting to climate risks (Section 3).

Explaining Responsible Business Conduct in Adaptation through Case Studies

To explore the second part of our theoretical framework laying out the factors that may explain RBC in adaptation, we conducted qualitative case studies on three mining projects in Peru. The selected mines in Peru include Antamina, Cerro Corona, and Quellaveco. Figure 1 indicates that these three projects are situated in Peruvian provinces, which are highly vulnerable to climate risks. Generally, Peru is not among the most vulnerable mining-dependent countries, but climate vulnerability is unevenly distributed across its provinces.

The case studies enabled a more nuanced and context-sensitive analysis of how each of the three factors shaped companies' integration of the

RBC principles. These mining companies rank among the most socially and environmentally responsible in the sector (RMI 2022). Studying how they implement the principles at the operational level enables us to identify important barriers to fostering greater RBC in adaptation (Section 4). We conduct the analysis of the three cases on the basis of material collected during four weeks of fieldwork conducted in the three mining localities and in the capital, Lima, in November and December 2022. We conducted forty-six interviews with representatives of the three mining companies, state agencies at both the national and subnational levels, and civil society and grass-roots organizations who work on climate adaptation, mining, or both (see Appendix C). We use these interviews to operationalize the three theorized explanatory factors and offer a context-sensitive analysis of the three cases.

To analyze *public governance* in Peru, we analyzed national-level legal requirements to integrate RBC into key sectoral policies, including EIAs, water licenses, the design of tailing dams, and closure plans. We also relied on the interviews to investigate subnational and national authorities' attempts to influence the companies through soft steering, such as partnerships, dialogues, and production of data on climate risks. Regarding the role of *voluntary adaptation standards* in shaping RBC in adaptation, we used both the quantitative dataset described earlier and interviews with company representatives to identify the most commonly used standards and then used documents describing the standards to analyze the extent to which these standards covered RBC principles. Subsequently, we interviewed representatives of the three companies to understand how they perceived the influence of these standards on their adaptation responses. Finally, we interviewed representatives from civil society and the three companies to investigate how *civil society advocacy* mattered RBC in adaptation. In summary, the case studies enabled us to systematically explore the theoretical framework in a context-sensitive manner.

1.3 Findings and Contributions

The central findings are twofold. First, both the quantitative and qualitative evidence suggest that companies have primarily addressed climate adaptation through conventional CSR, rather than more meaningfully incorporating other more procedurally oriented components of RBC. While 24 percent of the companies reported upon having made *CSR investments* in adaptation, 16 percent of the companies reported upon *assessing climate risks to society* and set up stakeholder participation processes in adaptation interventions. None of the companies lived up to the *transparency* principle and disclosed information to affected groups and governmental agencies about their adaption responses. Consequently, while there appears to be a group of sustainability frontrunners that have begun to integrate RBC principles into their adaptation strategies, most companies are found to continue to approach adaptation primarily as a matter of business resilience.

Second, based on the qualitative case studies from Peru, our findings underpin the theoretical expectations regarding the combined impact of public governance, private voluntary standards, and CSO advocacy on RBC in adaptation. Our analysis shows that these factors have primarily incentivized companies to make CSR investments, rather than pressuring them to integrate the procedurally oriented principles (risk assessments for society, stakeholder participation, and transparency). Regarding the influence of *public governance* on RBC, the Peruvian government has primarily encouraged companies to form partnerships to secure CSR investments in water infrastructure, rather than imposed legal requirements for companies to integrate the RBC principles in key sectoral policies.

Similarly, *civil society advocacy* has largely focused on demanding CSR investments to help affected communities to cope with the combined effects of mineral extraction and climate change. Companies in the three cases have adopted *voluntary adaptation standards*, which have contributed to spreading norms about RBC principles. However, the lack of strong enforcement mechanisms and in the absence of the countervailing power of public governance and civil society advocacy, these standards have had limited effects on company behavior. Taken together, we see some indications of an emerging discourse on RBC in adaptation. But companies have to date only made half-hearted attempts to change their behavior by more meaningfully incorporating the RBC principles in adaptation responses at an operational level.

Although these findings are based on analyses of the mining sector, we expect them to be generalizable mainly to two contexts. First, our findings are particularly relevant for land- and water-intensive sectors, such as energy, food, oil, gas, hydropower, and large-scale agriculture. In these sectors, corporate adaptation is likely to influence local ecosystems and communities. The findings may also be relevant for working conditions in other industries increasingly exposed to heat waves, where climate adaptation is i recognized as essential for protecting workers (Judd et al. 2023). Second, our findings are also likely to apply to climate-vulnerable countries reliant on key commodities, where economic dependence may make governments hesitant to enforce stricter regulatory controls on businesses.

With these considerations in mind, our findings contribute to two ongoing debates related to corporate adaptation governance and public–private sector interactions, as discussed at length in Section 5. First, they engage with an emerging and growing literature on the role of the business actors in adaptation (e.g., Averchenkova et al. 2016; Berkhout 2012; Gasbarro and Pinkse 2016; Goldstein et al. 2019; Grabs et al. 2026; Gustafsson et al. 2022; Linnenluecke et al. 2011; Mbanyele and Muchenje 2022; Shi and Moser 2021). While several of these studies have concluded that companies rarely take societal impacts into account when addressing climate risks (e.g., Goldstein et al. 2019; Shi and Moser 2021), this behavior is, as Averchenkova and colleagues (2016) point out, an under-researched topic. This study advances emerging debates about the role of companies in adaptation governance by conceptualizing, explaining, and providing for the first systematic analysis of RBC in the context of adaptation.

Second, this study offers insights into the relationship between businesses and the state in the area of adaptation. While a substantial literature has analyzed private–public sector relations in sustainability governance more broadly (Cashore et al. 2021; Eberlein 2019; Lambin and Thorlakson 2018), there are relatively few studies focused on adaptation (notable exceptions are Klein et al. 2018; Pauw and Pegels 2013). Our findings reveal that, to date, Peruvian public authorities have relied on soft steering to incentivize companies to fund adaptation infrastructure, rather than imposing regulatory controls. However, without strict legal requirements, such soft regulatory approaches have proven ineffective in promoting significant change in corporate behavior.

1.4 The Structure of the Element

This Element is structured as follows. Section 2 introduces our theoretical framework on RBC in climate adaptation. Section 3 explores the first part of this framework, mapping how the world's largest mining companies understand climate risks, engage in adaptation, and integrate the four RBC principles, based on our quantitative database. In turn, Section 4 turns to the second part of our framework and analyzes how regulatory and political factors shape RBC in adaptation based on the case studies in Peru. Section 5 summarizes our key findings, discusses those findings against the backdrop of previous research, and outlines their broader implications for the future of climate adaptation governance.

2 A Theoretical Framework for Studying Responsible Business Conduct in Adaptation

As Section 1 established, gaining a deeper understanding of how firms approach RBC principles in adaptation is critical in the context of increasingly severe climate impacts. In this section, we provide a detailed account of our two-part theoretical framework, which was briefly introduced in Section 1. In the first part (Section 2.1), we conceptualize the principles, which cover both procedural and distributive aspects of RBC. Drawing on literatures on private governance, Business and Human Rights, and climate adaptation, we discuss how these general principles manifest in the adaptation context and why we regard them as critical. In the second part (Section 2.2), we discuss different regulatory and political factors which are likely to shape RBC in the context of adaptation. We zoom in on public governance, private voluntary standards, and civil society advocacy and discuss how each is likely to influence business behavior.

2.1 Corporate Responsibility Principles in Climate Adaptation

We define RBC as the behaviors of companies which avoid and address the negative impacts of their activities, while also fostering sustainable development in the countries where they have operations (cf. OECD n.d.). Although it is a central concept in the literatures on CSR and Business and Human Rights, it has not yet been transposed to the context of climate adaptation. The RBC principles we elaborate on in the following section cover the procedural and distributive dimensions that enable us to go beyond the "do no harm" approach of proceduralist approaches and include principles which could potentially capture distributive outcomes. We do not claim that this constitutes a comprehensive overview of all RBC principles embedded in international soft and hard law. For example, we do not include corporate governance and access to remedy, both of which are central components of HREDD. However, we argue that the selected principles together capture the initial expressions of central principles that are particularly relevant for preventing socio-environmental harm and for developing regulatory policies addressing the intersecting impacts of climate change and business activities on society.

Assessments of Climate Risks to Society

Assessing climate risks to society enables companies to identify and address potential harms linked to their operations, making it an important procedural principle for preventing harms before they occur. We understand assessments of climate risks to society as the obligation of companies to adequately evaluate the societal risks associated with their business activities in a context of climate change. Climate risk assessments involve a technical analysis of the likelihood, responses, and consequences tied to climatic impacts, and the options available to redress them (see also Adger et al. 2018). While many companies have procedures to assess their own risk exposure, including those ensuing from climate impacts, addressing socio-environmental risks requires shifting the focus toward the external risks posed by their own activities.

To gather sufficiently rigorous data, companies must conduct a complex technical analysis that will yield reliable data showing how climate impacts could exacerbate, or even create, new socio-environmental risks linked to their business activities. In the mining sector, companies must ensure that tailings dams withstand intense rains (Hopkins and Kemp 2021). In the agroindustry, companies also must assess the short- and long-term effects of their water use (cf. Rattis et al. 2021). Meanwhile, in the apparel industry, the effects of extreme heat and floodings on workers' health and safety must be carefully analyzed (Judd et al. 2023).

To carry out climate risk assessments, companies can integrate climate risks into existing forms of risk assessments. For example, in the case of major projects, companies in most countries are legally required to conduct EIAs and are thereby already assessing their potential impacts on the environment and, in some cases, on society (Lawrence and Kløcker Larsen 2017). Moreover, HREDD frameworks give companies tools to ascertain the human rights and environmental risks associated with their business activities, though climate change, and environmental issues more broadly, have often been poorly integrated in such procedures (Bernaz 2016; Macchi 2022). Hence, companies could integrate climate adaptation into existing methods of risk assessment rather than developing new ones.

Despite the potential to integrate climate adaptation into existing risk management procedures, barriers hinder companies from carrying out such assessments. One major challenge lies in the complexity and uncertainty of these assessments, which can reduce companies' motivation to conduct them. Climate impacts can have immediate, short-term, and long-term consequences. In some cases, the impacts do not arise until after a company has ceased its operations. In addition, companies must contend with the uncertainty surrounding future climate projections, specifically in assessing long-term impacts and the effectiveness of adaptation interventions (Adger et al. 2018). Research shows that perceived uncertainty influences how much attention companies allocate to climate impacts and whether they adopt more rigorous adaptation responses (Pinkse and Gasbarro 2019).

In sum, assessing climate risks to society is a procedural principle that enables companies to address harms before they occur and can often be integrated into existing risk assessment procedures. But significant challenges remain due to the complexity of analyzing how business activities might interact with climate change. That these assessments also support other procedural principles such as participation and transparency makes them even more essential for RBC in adaptation.

Participation

Meaningful participation ensures that affected groups can influence how their perspectives and adaptation needs are integrated into corporate adaptation responses. Specifically, we refer to participation as the engagement of relevant stakeholders in the planning and implementation of adaptation interventions (Eriksen et al. 2021; Shi et al. 2016). According to the UNGPs (OHCHR 2011) and the OECD Guidelines (2023), companies shall consult stakeholders throughout the process of assessing and addressing environmental and human rights risks. Companies can consult affected stakeholders regarding climate risk assessments and the implementation of adaptation activities. This is especially important when corporate adaptation efforts impact the land and water that local communities rely on for their livelihoods (cf. OECD 2023: 20).

Depending on their purpose, quality, and outcomes, participatory processes may give vulnerable and marginalized groups greater voice and influence (Dellmuth and Gustafsson 2023; Gustafsson and Schilling-Vacaflor 2022). To foster meaningful stakeholder consultations, the OECD Guidelines stipulate that companies shall set up participatory processes that are "timely, accessible, appropriate and safe for stakeholders" (OECD 2023: 20). They further underscore the importance of identifying and removing potential barriers to engagement with vulnerable and marginalized groups (OECD 2023: 19).

However, participatory processes have often not enabled the influential participation of marginalized groups. For example, adaptation interventions led by public actors have often suffered from shortcomings when it comes to setting up inclusionary participatory processes (Eriksen et al. 2021). Similarly, company-led participatory processes in large-scale projects are often beset by shortcomings such as limited information, time, and influence on decision-making (Lawrence and Kløcker Larsen 2017). Consequently, if companies formally involve affected groups in implementing adaptation interventions but fail to provide adequate and accessible information about climate risks, they undermine their ability to voice their concerns and influence corporate adaptation strategies.

Transparency

As the final procedural principle, access to information on climate risk assessments and corporate adaptation responses equips affected groups and public authorities with tools to critically evaluate how companies address climate-related risks arising from their operations. We conceive of transparency as companies' duty to inform society more broadly about the societal risks associated with their business activities in the context of climate change and of the adaptive measures they have taken to address them (cf. OECD 2023). Access to such information is particularly important in cases where business activities may expose communities and ecosystems to severe risks. For instance, if a company builds a pipeline to transport toxic materials, it must inform local communities of any increased risks for dangerous accidents.

There has been a proliferation of voluntary initiatives and legislation mandating that companies report on their sustainability performance, particularly regarding climate change. Initiatives like the Carbon Disclosure Project and the Global Reporting Initiative have significantly enhanced companies' reporting about their decarbonization strategies (cf. Pinkse and Gasbarro 2019). In contrast, few initiatives require companies to disclose information about their adaptation strategies.

Corporations tend to be very cautious in disclosing information about their adaptation interventions (Gustafsson et al. 2022). Unless required to do so, those operating in high-risk sectors, such as extractive industries or agroindustry, generally have few incentives to be transparent about the risks they face, as doing so could provoke conflicts that disrupt their operations. This situation is very different in the climate mitigation space, where companies are strongly motivated to showcase their contributions to the low-carbon transition (Jernnäs and Lövbrand 2022).

Although climate risk assessments are technically complex and their results uncertain, companies must find a way to present the information they gather in a format that is relevant and easily accessible to affected groups (cf. Gardner et al. 2019; OECD 2023). Without such information, local populations and public institutions will struggle to evaluate corporate adaptation responses. In turn, this may undermine local and public actors' attempts to hold companies accountable and hinder the establishment of a democratic dialogue on the impacts of business activities within the context of climate change.

CSR Investments

CSR investments can enable companies to move beyond a "do no harm" approach and make positive contributions to adaptation in the countries where they operate. In our framework, CSR investments are understood as the provision of public adaptation goods by companies (cf. Tompkins and Eakin 2012). Companies can invest in early warning systems, cooling systems, water infrastructure, and capacity building to help affected stakeholders to cope with climate change. Such investments could expand access to public adaptation goods in countries where governments lack the resources to fund the implementation of adaptation plans. The problem is particularly urgent in developing countries, where the annual adaptation costs are expected to reach 140 to 300 billion USD in 2030 and 280 to 500 billion USD in 2050 (UNEP 2021: xiv). According to the UN Environment Programme (UNEP), private adaptation funding is essential to reducing this gap (UNEP 2021: xiv).

Against this background, companies could partner with nongovernmental organizations (NGOs) and state institutions to invest in essential public adaptation goods such as water infrastructure, climate-proofing infrastructure, and capacity building. By channeling CSR investments through institutionalized partnerships that benefit broad swaths of the populations, companies could contribute both through financial support and by expanding state capacity to implement adaptation plans.

At the same time, companies are often unwilling to make CSR investments unless there is a business case (Auld et al. 2008). CSR investments in the mining sector tend to be aimed at preventing or mitigating conflicts with local communities that could disrupt their economic activities (Franks et al. 2014). To defuse opposition, companies sometimes make targeted investments that benefit smaller groups, but create dependency which often undermines long-term local development, civil society organizations, and institutional reforms (Amengual 2018; Bebbington 2010; Gustafsson 2018). In such situations, the existence of CSR investment might reduce local communities' demands for the procedurally oriented principles. Consequently, it is important to monitor how companies channel these investments, be it either through institutionalized partnerships or bilateral exchanges with small groups or individuals as tools for gaining political support.

2.2 Explaining Corporate Responsibility in Climate Adaptation

Next, having laid out the first part of our theoretical framework, we turn to the second part, where we explain how the three factors external to the firm – public governance, private voluntary standards, and civil society advocacy – shape RBC in adaptation. These factors have been shown to affect corporate behavior more broadly (Bartley 2018a; Lambin and Thorlakson 2018; Thorlakson et al. 2018a), underlining their potential relevance in the specific context of adaptation. While previous studies have analyzed how companies' exposure to physical climate change (Mbanyele and Muchenje 2022) and internal perceptions of these risks (Berkhout 2012; Pinkse and Gasbarro 2019; see Gasbarro and Pinkse 2016) influence corporate adaptation, we focus on how these regulatory and political factors shape RBC in adaptation.

Public Governance

We expect public governance to influence corporate behavior through both legal requirements and soft steering. Public governance refers to the system of organizations, institutions, and networks at domestic and global levels that develop and implement rules and norms for handling societal problems (Weiss and Wilkinson 2019). We analyze how domestic institutions at both the national and subnational levels require or encourage RBC in adaptation through hard and soft regulatory activities. These forms of governance are important because they can strongly influence corporate behavior by shaping incentives and enforcing compliance.

Legal requirements are central to influence business behavior. For instance, governments can institute legal requirements that climate risks be considered within key sectoral regulatory instruments such as concessions, water licenses, and EIAs. These instruments can act as "market-restricting rules" that governments could employ to limit or prohibit business activities likely to undermine societal and ecosystemic adaptation (cf. Bartley 2018a). In developed countries, the prospect of governmental regulation has been an important driver of corporate climate action, particularly in the area of mitigation (Newell and Daley 2024; Sakhel 2017). However, legal requirements in the area of adaptation tend to be weaker (Klein et al. 2018), especially in countries in the Global South (Pulver and Benney 2013; Purdon and Thornton 2019).

Still, most countries have EIA legislation in place, which can serve as an important instrument for governments to require companies to conduct impact assessments, disclose information, and enable stakeholder participation. Furthermore, EIAs could lead to stricter rules or the suspension of major projects where the negative impacts are deemed too high. While EIA legislation increasingly includes climate mitigation, legal requirements that climate adaptation be taken into account are rare, reflecting that adaptation is not yet institutionalized in sectoral policies (Mayembe et al. 2023). However, some governments have also used EIAs as an instrument to restrict activities deemed to have a negative impact on climate adaptation. For example, in 2017, the Salvadoran government conducted a Strategic Environmental Impact Assessment to evaluate the impacts of mining under conditions of increased water scarcity and climate-related disaster. That assessment resulted in a national ban on metal mining (Odell et al. 2018).

Governments can also use support and soft steering mechanisms, like funding, scientific information, endorsement, partnership, or multistakeholder initiatives to influence companies' adaptation behavior (cf. Abbott and Snidal 2009; Eberlein 2019). For example, some countries have developed voluntary guidelines on how companies should assess physical climate risks in EIA processes (Strindevall et al. 2022). Since knowledge and expertise are often important barriers to tackling climate impacts, clear guidelines can encourage companies to undertake risk assessments for society by making them less costly. Moreover, partnerships and multistakeholder initiatives can provide companies with institutionalized channels through which they can channel CSR investment in public adaptation goods, disclose information, and engage important stakeholders in dialogues on adaptation. However, soft steering processes are often ineffective in changing company behavior unless backed up by a viable threat of mandatory regulation (Börzel et al. 2012).

To effectively steer or regulate company behavior, public institutions must involve robust monitoring systems capable of assessing the impacts of climate change and evaluating the extent to which companies responsibly assess and redress manifest risks (cf. Lambin and Thorlakson 2018). While many governments have adopted National Adaptation Plans (NAPs), the quality of the data that emerges is a significant limitation. NAPs are often too generic to capture risks in specific geographic and social contexts (Dolšak and Prakash 2018). Without robust data on climate risks, public authorities cannot provide companies with relevant data or evaluate the adequacy of corporate responses.

Adaptation is often not prioritized by political and economic actors. Where national or subnational economies heavily depend on particular economic activities, the adaptation agenda may become more responsive to corporate interests than to the rights and needs of marginalized groups. In Brazil, for instance, adaptation policies have enabled large-scale agricultural producers to cope with water stress through irrigation infrastructure, often at the expense of marginalized communities who face severe water insecurity as a result (Gustafsson et al. 2024; Milhorance et al. 2022a). Overall, it is important to consider the underlying political priorities, power dynamics, and monitoring capacities that shape how legal requirements and soft steering influence company behavior.

Voluntary Adaptation Standards

We expect voluntary adaptation standards to matter for RBC in adaptation through the diffusion of norms about RBC. Voluntary adaptation standards refer to any type of multistakeholder initiative, certification scheme, or private standard that sets requirements designed to increase the sustainability of corporate adaptation responses (cf. Marx et al. 2024). More generally, voluntary sustainability standards typically define specific foundational principles, which are then translated into measurable indicators for monitoring and compliance (Marx et al. 2024: 3). However, standards differ in terms of their institutional design, and although they usually rely on enforcement mechanisms such as audits, they can also rely on market incentives and capacity building to generate compliance (Depoorter and Marx 2024). When public governance is weak or lacking, companies are more likely to rely on voluntary standards (cf. Potoski and Prakash 2013). This is likely to be the case in adaptation.

While there are no comprehensive voluntary adaptation standards, several standards have integrated climate adaptation to some extent, notably those developed by the Alliance for Water Stewardship, the Sustainable Agriculture Initiative, and the Responsible Business Initiative. In the mining sector, ICMM and the *Global Tailings Review* have developed industry-specific standards which include adaptation (Hopkins and Kemp 2021). Such standards can contribute to learning and thereby facilitate businesses uptake of RBC principles (cf. Depoorter and Marx 2024).

The private governance literature yields mixed evidence about whether voluntary standards lead to meaningful improvements in the sustainability practices of firms (Marx et al. 2024; Thorlakson et al. 2018a). A study by Thompson and colleagues (2021) on the effects of certifications on the climate resilience of cocoa farmers in Ghana showed some positive impacts on farm management practices, but found significantly weaker effects on more complex adaptation measure. This is likely due to the continued lack of clarity regarding what constitutes adaptation, which makes it challenging for standard setters to develop and monitor appropriate measures. More generally, numerous studies have shown that voluntary standards are often insufficient for preventing or rectifying serious problems due to limited uptake, lack of stringency, poor-quality auditing, and a tendency toward symbolic compliance (Bartley 2018b; LeBaron et al. 2017).

Companies usually adopt voluntary standards for utilitarian reasons (de Bakker et al. 2019). Retailers with high-profile consumer brands, who are more exposed to CSO scrutiny, have shown a greater tendency to adopt sustainability standards compared to business-to-business firms (Thorlakson et al. 2018a). Additionally, researchers have concluded that the use of public monitoring systems and legal enforcement mechanisms contribute to more meaningful compliance with voluntary standards (Lambin and Thorlakson 2018). In sum, while voluntary standards can contribute to learning about RBC in adaptation, meaningful adoption and compliance by profit-maximizing firms tends to depend on incentives or external pressures, with civil society advocacy often playing an important role.

Civil Society Advocacy

Finally, we expect CSO advocacy to influence RBC not only by pressuring them to adopt voluntary adaptation standards but also to develop more inclusionary and transparent adaptation responses.

Here, civil society advocacy refers to a civil society-based form of corporate regulation (cf. Newell 2008), in which NGOs, social movements, grassroots organizations, labor unions, and other civil society actors pressurize companies to change their behavior (cf. Dellmuth and Bloodgood 2023). CSOs have often played a critical role in scrutinizing companies and incentivizing companies to disclose information, to reduce the negative impacts of their business activities, to invest in CSR, to adopt voluntary sustainability standards (Bartley 2018b; Franks et al. 2014; Thorlakson et al. 2018a), and to comply with legal requirements (Kramarz et al. 2023).

CSOs pursue a range of advocacy strategies, such as generating and disseminating alternative data about the impacts of business activities (Kramarz et al. 2023), engaging in highly visible naming-and-shaming campaigns (Balsiger 2014), and interacting closely with companies through multistakeholder initiatives (Gustafsson 2018). In particular, the deployment of visible campaigns could pressure companies to integrate RBC. Campaigns that show the consequences of failing to address climate risks can lead to consumer boycotts, investor demands for action, and other potential consequences that reputation-sensitive businesses usually seek to avoid (Balsiger 2014). Concerns about reputational harm might lead companies to improve climate risk assessments, develop their adaptation responses in a more transparent and inclusionary way, and support societal adaptation through CSR investments.

However, civil society advocacy in the field of climate adaptation has been relatively weak, in part because adaptation is often framed as a technical and depoliticized problem. In sharp contrast to civil society campaigns against carbon-intensive industries (Boulianne et al. 2018), activists have rarely engaged in open protests to pressure governments and corporations to intensify their efforts to promote climate adaptation (de Moor 2022). Social movement scholars have stressed that CSO mobilization around systemic issues, where responsibility is diffuse and specific attribution challenging, is generally less likely to occur (Carpenter 2007). This is certainly the case with corporate adaptation, where it can be difficult to provide robust evidence on a given corporation's impact on community and ecosystem resilience. At the same time, local protests over climate-related impacts are common – especially in industries that require the intensive use of land and water – even if they are not explicitly framed in terms of climate change (Leme da Silva et al. 2023; Odell 2021; Urkidi 2010).

Overall, while CSO advocacy has the potential to pressure companies to integrate RBC principles into their adaptation responses, its role in shaping corporate behavior depends on the ability of civil society actors to mobilize collectively and articulate concrete, actionable demands regarding corporate responsibilities in the face of climate risks.

2.3 Conclusion

This section has developed a theoretical framework for understanding the main outcome of interest, RBC in the context of climate adaptation. Drawing on literatures on private governance, Business and Human Rights, and climate adaptation governance, we proceeded in two steps. First, we conceptualized four RBC principles in adaptation, each of which is essential for preventing and mitigating the socio-environmental harms associated with business activities under climate change and foster more coordinated private and public governance of such impacts. To move beyond a "do no harm" approach, we also included one distributive principle: CSR investments in public adaptation goods. Second, we developed an argument about three regulatory and political factors that we expect to explain how companies integrate these principles into their adaptation responses. Our explanation privileges public governance, voluntary adaptation standards, and civil society advocacy.

The empirical analysis in the next section will draw on the first part of this framework, while the subsequent section will apply the full framework.

3 Mapping Responsible Business Conduct in Adaptation

This section explores the first part of our theoretical framework by descriptively mapping the extent to which the world's largest mining companies have integrated the RBC principles into their adaptation responses. For this purpose, we draw on the dataset introduced in Section 1, which is based on a content analysis of corporate documents from the world's thirty-seven largest mining companies and on semi-structured interviews.

To set the stage for the descriptive mapping, we begin by providing contextual information about mining companies' perceptions of climate risks and their organizational responses to adaptation problems (Section 3.1). We then proceed to mapping the variation in the main outcome of our theoretical framework, RBC in adaptation, operationalized as the extent to which firms adhere to the four conceptualized principles for RBC in adaptation (Section 3.2). The section concludes with a brief summary and a discussion of the evidence in light of our argument (Section 3.3).

3.1 Firms' Perceptions of Adaptation Problems

Before mapping companies' adherence to the RBC principles, we begin by describing how mining companies make sense of climate risks in the first place. More specifically, we analyze whether companies perceive climate change as an economic risk or as one with implications for both the economy and society. Relying on the notion of "risk" makes sense here, as companies' conventional methods of risk management are typically focused on their own exposure; risks for business resilience tend to be incorporated into a company's risk management framework (Higham 2019). In contrast, approaching climate risks from a societal perspective reflects a more comprehensive and context-sensitive understanding of how biophysical processes and economic activities interact to produce societal climate vulnerability (cf. Bassett and Fogelman 2013; Odell et al. 2018).

Climate adaptation, for the majority of companies in our sample, is at least not primarily aimed at mitigating societal climate vulnerability, but a strategic response to the economic risks it poses to their operations and long-term profitability. Figure 2 shows that among the thirty-seven companies in our sample, 70.3 percent portrayed climate impacts as an economic risk to their business operations ($N = 26$). Company reports describe how floodings, increased storm intensity, and high temperatures could damage

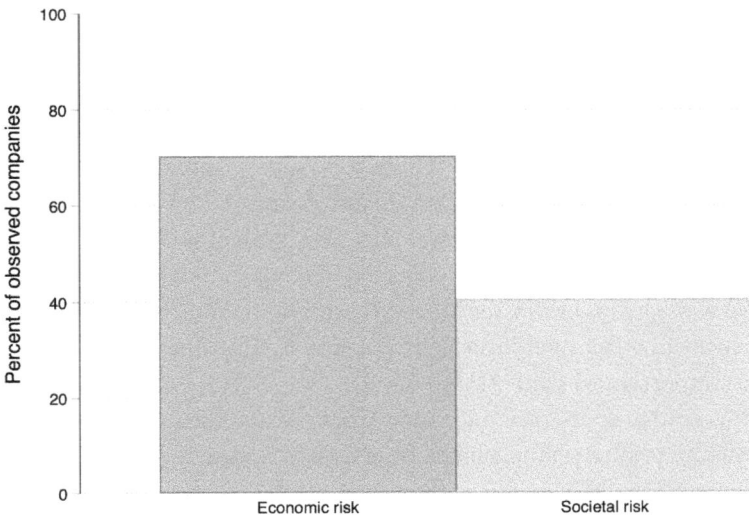

Figure 2 Company perceptions of adaptation problems.

Notes: $N = 37$ companies. Pooled data for the period 2017–2019 from company documents. The indicators were coded Yes = 1/No = 0 for each company based on the following questions. *Economic risk*: Do companies recognize climate change as a risk for business performance and continuity? *Societal risk*: Do companies recognize the physical impacts of climate change as a societal risk? See Appendix B (Table B1) for the detailed coding scheme.

mining infrastructure, exacerbate water stress, disrupt production processes, and result in costly litigation. However, only half of the companies identified climate impacts as a "critical" or "principal risk." The extent to which companies prioritize adaptation is also shaped by the extent of their risk exposure. For instance, some, like Antofagasta, have all their operations in water-stressed areas, while others, like Anglo American, have a majority of their operations in such regions (Anglo American 2019; Antofagasta 2019).

Interestingly, companies' perceptions of economic risks differ significantly between adaptation and mitigation, suggesting that adaptation is not the main priority for many companies. Our document analysis reveals that a majority of the companies recognize that the failure to reduce their own greenhouse gas emissions is a critical risk likely to have significant financial impacts. Yet concerns of this kind appear less salient in the adaptation context. Interviewees in the companies highlighted several reasons for mining companies' seeming reluctance to recognize climate impacts as a critical risk for their own operations, including the inability to accurately assess the magnitude of future climate-driven impacts, the cost of adaptation, and the absence of strict regulatory controls (Interviews 9, 10, 45).

Some mining companies have begun to frame climate change as a societal risk, recognizing that harm to surrounding communities can lead to reputational damage for the company. Returning to Figure 2, our research demonstrates that 40.5 percent of the thirty-seven companies see climate impacts as a *societal risk* ($N = 15$). These companies expressed particular concern regarding the possibility that climate change could lead to water stress and tailings dam failures, events that could affect their reputations and lead to local protests. As one company representative stated: "If we don't have water available for the affected communities and we don't meet our commitments, the operation will shut down. Just like that. I mean water is as important as gold" (Interview 2).

To address the economic and social risks associated with climate change, effective water management is generally considered a key adaptation intervention (ICMM 2017; Interviews 2, 41). Water is one of the most important drivers of community protests against mining projects. Several mining-dependent countries, most prominently Bolivia, Chile, the Democratic Republic of Congo, Mongolia, Namibia, Peru, and South Africa, face high or extreme levels of water stress (Strindevall et al. 2022). Moreover, intense rainfalls can lead to dangerous accidents and severely harm surrounding communities and environments. A case in point is the catastrophic failure of Vale's tailings dam in Brumadinho, Brazil, in 2019, which cost over 250 people their lives and caused significant material destruction (Hopkins and Kemp 2021).

In sum, while most mining companies perceive climate impacts primarily as an economic risk to their operations, far fewer recognize the broader societal dimensions of climate vulnerability. Water stress, in particular, emerges as a critical issue – both as a driver of community conflict and as a fundamental challenge to long-term operational resilience.

Next, we deem it necessary to analyze how companies have responded to climate risks, before turning to the main outcome of interest, how they considered the *societal consequences* of their business activities in a context of climate change. In our document analysis, we inductively identified four broad organizational responses to climate risks: water governance, risk management, investments to climate-proof infrastructure, and mine closure. Figure 3 shows that 54.1 percent of companies integrated climate risks into water governance ($N = 20$), 48.4 percent incorporated climate risks into risk management ($N = 18$), 46.0 percent made investments to climate-proof their infrastructure ($N = 17$), and 13.5 percent sought to redress climate risks in their plans for mine closures ($N = 5$).

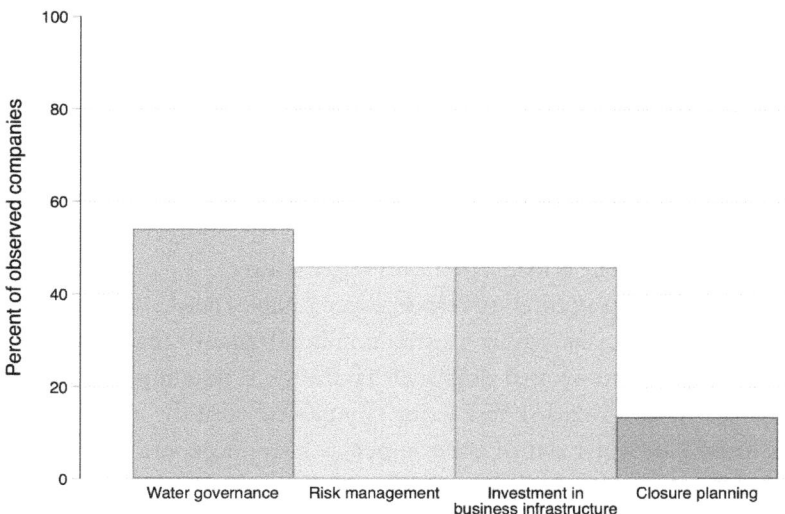

Figure 3 Company responses to adaptation problems.

Notes: $N = 37$ companies. Pooled data for the period 2017–2019 from company documents. The indicators were coded Yes = 1/No = 0 for each company based on the following questions: Water governance: Do companies integrate climate risks in water governance? Risk management: Do companies integrate climate risks in their procedures for risk assessment and management and business plans? Infrastructure investment: Do companies invest in climate-proofing their own infrastructure? Closure planning: Do companies consider climate change in their mine closure plans? See Appendix B (Table B2) for the coding scheme.

In the following, we briefly discuss each of these four organizational responses, beginning with the most frequently observed type of response: water governance. First, organizational responses to water stress represent an important adaptation response in the mining sector, with 54.1 percent of companies taking steps to monitor and set targets for reducing their water consumption. Several of these companies report using screening tools developed by NGOs, such as the World Wildlife Fund for Nature and the World Resources Institute, to make initial assessments and identify areas under severe water stress. Based on these screening tools, companies have developed water balance models that consider climate impacts, set targets to reduce their consumption of freshwater, and increase water recycling or use of seawater in water-stressed areas. However, an interviewee experienced with global-level screening pointed out the need to complement this information with more granular local data on water resources and usage (Interview 87). Indeed, a water expert working at the United

Nations Development Programme (UNDP) office in Peru emphasized that it is almost impossible to discern the water footprint of mining companies from the outside, as they do not disclose the indicators they use. Neither are companies inclined to be transparent about water pollution (Interview 31), which is a severe and longstanding problem for local communities (Lèbre et al. 2020). Consequently, it is difficult to assess whether the measures taken to improve water governance adequately address the climate change-related challenges associated with water stress.

Second, to develop adaptation responses, companies need a solid understanding of climate risks. Mineral extraction is extremely sensitive to different types of climate-related risks such as droughts, flooding, and rising temperatures (Phillips 2016). Reflecting companies' need for understanding such risks, 48.4 percent of the companies have incorporated climate impacts into their conventional risk assessment and management frameworks. All but one of them report having conducted site-based climate risk assessments and integrated climate risks in their risk management procedures. Most companies have developed internal capacities within their environmental units to monitor climate risks, sometimes in partnership with external consultants. Several interviewees working with climate risk assessments within the mining companies highlighted that a major challenge was combining the IPCC's five global climate scenarios with their own local data to make reliable predictions about short- and long-term climate impacts (Interviews 6, 7, 36).

All companies that reported integrating climate adaptation into their risk management procedures stated that their board had oversight of climate-related risks, making it almost certain that the topic is discussed at senior management levels and within the board. This not only indicates the companies' recognition of the problem's importance but also that climate-related risks are likely to be integrated into business strategies, even if our data do not reveal how much priority companies actually give to climate adaptation.

Third, in response to the growing impact of droughts and extreme weather events on mining operations, 46.0 percent of the companies reported making investments to climate-proof their infrastructure. The most common investments were in measures to ensure that tailings dams could withstand intense rainfalls, and technological adjustments and water storage infrastructure to cope with water stress. While these companies often provided specific examples of technical solutions to climate-proof their infrastructures, they tended not to disclose the total amount spent on these investments. Newmont is an exception. The company reported that it cost "tens of millions of dollars" to update its infrastructure over the prior three years (Newmont 2019: 34). However, this information cannot be readily verified, and it is impossible to determine the significance of the investment in relation to the identified climate risks.

Fourth and finally, after mining activities have ended, they often leave behind dangerous waste that can pollute land and water, making mine closure a critical process for preventing long-term environmental harm – especially as future climate impacts like heavy rainfall or flooding could make these risks even worse (Macklin et al. 2023). Yet only 13.5 percent of companies reported accounting for climate change in their closure planning. Failing to do so could expose surrounding communities and ecosystems to serious and lasting harm.

This variation in adaptation responses can be better understood when considering the role of ICMM. As an important industry initiative, ICMM has acted as a standard-setter by including adaptation in its Mining Principles, with which member companies are required to comply. Notably, thirteen out of twenty-one companies in our sample that have developed organizational responses are members of ICMM. These companies have collaborated with the organization to develop sector-specific guidelines, tools, and best practices for assessing climate risks (ICMM 2016, 2021). ICMM member companies appear to be more advanced in their adaptation efforts, often providing more detailed and structured information about their practices than nonmembers. For example, Anglo American, Rio Tinto, South 32, and Teck describe having relatively advanced risk assessments and technical solutions in place to address climate impacts. Rio Tinto has worked on climate-related risks since 2003 (Rio Tinto 2018), and Newmont began implementing its adaptation program in 2016 (Newmont 2019). However, not all ICMM members view adaptation as a priority. AngloGold Ashanti, for example, do not consider it a critical risk and continue to focus mainly on mitigation (AngloGold Ashanti 2020; Interview 9). We explore ICMM's role more fully in Section 4.

Despite operating in countries highly vulnerable to climate change, many mining companies in our sample have not yet developed any organizational response to address climate risks. Sixteen companies – including several based in China, Russia, India, and Saudi Arabia – do not report upon having any adaptation measures in place. Notably, three of these operate in the Democratic Republic of Congo, one of the world's most climate-vulnerable countries (ND-GAIN 2018). This lack of action could have serious societal consequences, highlighting the importance of RBC in adaptation – our main outcome of interest to which we now turn.

3.2 Responsible Business Conduct in Adaptation

By exploring the extent to which companies have integrated the four principles of RBC in adaptation, as conceptualized in Section 2, the ensuing analysis maps the main outcome in our theoretical framework. In our document analysis, we have coded companies as adhering to an RBC principle if we can identify a general strategy or procedure for their integration. In this case, a company takes on the value of 1. We code companies as 0 if they referred to isolated and unsustained initiatives that provide no clear indication of having been systematically integrated into a cohesive approach to climate adaptation. It was only with this fine-grained coding that we were successful in capturing this variation.

Our document analysis reveals that 29.7 percent ($N = 11$) of the companies in our sample claimed to adhere to one or more of the four RBC principles. The companies comprising this group are: Anglo American, Antofagasta, Barrick, BHP Group, Freeport McMoran, Fresnillo, Grupo Mexico, Newmont, Rio Tinto, South 32, and Teck Resources.

We now discuss each RBC principle in the order of the frequency of their integration by firms. Figure 4 shows that 24.3 percent of companies reported CSR investments in adaptation ($N = 9$), followed by 16.2 percent that assessed the climate risks exposure of local communities ($N = 6$), and 16.2 percent that adhered to the participation principle ($N = 6$). None observed transparency in their adaptation activities ($N = 0$).

We begin with *CSR investments*, which we defined earlier as companies' provision of public adaptation goods that support climate resilience. Such initiatives can contribute to expanding adaptation infrastructure, but they also serve to sustaining corporate support and managing conflict, particularly in contexts of climate-induced water stress.

Turning to the first bar in Figure 4, 24.3 percent of the companies in our sample reported making CSR investments in public adaptation goods, most commonly in water-storage infrastructure. Companies described these projects as a way to enhance the water security of local communities impacted by the combined effects of climate change and mining activities. Several company representatives explained to us that an important rationale behind water storage projects was to compensate communities for the water actually used by the mine, thereby preempting or reducing community protests (Interviews 2, 3, 9). This suggests that companies often use CSR investments as a strategic tool for managing company–community relations in the context of climate change (cf. Franks et al. 2014).

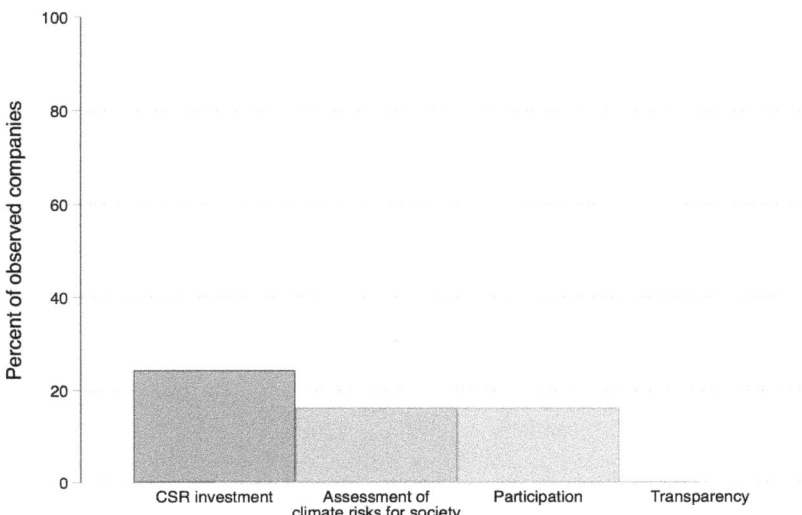

Figure 4 Percentage of companies adhering to principles of Responsible Business Conduct.

Notes: $N = 37$. Pooled data for the period 2017–2019 from company documents. The indicators were coded Yes = 1/No = 0 for each company based on the following questions. Assessments of climate risks for local communities: Do companies report upon assessing climate risks for local communities? CSR investment: Do companies report upon initiatives that they have developed or funded that are mainly aimed at providing public adaptation goods? Participation: Do companies report about having procedures in place for involving relevant stakeholders in the planning and implementation of adaptation interventions? Transparency: Do companies report on having procedures in place to inform local and/or national stakeholders about their exposure to climate risks and measures adopted to redress them? See Appendix B (Table B3) for the coding scheme.

As a result, companies have rarely invested in adaptation projects that do not directly contribute to maintaining community support. For example, only a few of the observed companies have supported capacity-building initiatives aimed at raising climate awareness or developing emergency plans for extreme weather events. This is notable, as generating awareness about climate adaptation among host communities is an explicit goal of ICMM. ICMM expects all member companies to be in compliance with this goal by the end of 2023 (ICMM 2021: 3). Yet, only a few companies in our sample have developed projects with the explicit goal of raising awareness of climate risks within local communities. In 2016, Rio Tinto launched a pilot project in Peru to enhance communities' capacity "to understand, prepare for, and respond to the impacts of climate change"

(Rio Tinto 2016: 4). The company planned the project as a model that other ICMM members could replicate. Gold Fields implemented a similar project in Peru between 2018 and 2020 (Lutheran World Relief 2017). However, none of the companies continued to implement these projects. Interviewees involved in these projects explained that the companies abandoned the projects in part because community members were more focused on immediate concerns, such as employment and compensation, and some were suspicious of companies' motives to talk about climate impacts. In this context, capacity-building initiatives aimed at raising climate awareness have struggled to gain traction or build trust among local communities, offering companies little incentive to continue or expand such initiatives (Interviews 5, 8, 37, 78).

Assessing climate risks to society is a crucial procedural step for preventing harm, yet our findings show that most companies remain in the early stages of evaluating how their operations may contribute to societal vulnerability in the context of climate change. As illustrated by the second bar in Figure 4, 16.2 percent of the companies reported having analyzed the vulnerability of host communities in their site-based risk assessments. However, until 2019, not a single one of the companies studied provided information describing the procedures they used in conducting their vulnerability assessments; neither did they clearly distinguish between their own risk exposure and that of the participating communities. A company representative put it bluntly: "Even if we look very selfishly at our own [climate] risks, it will still have positive impact on the community because it reduces the [community] risk" (Interview 41). These findings underscore that while companies primarily focus on assessing their own climate risk exposure, their community climate risk assessments remain underdeveloped.

While a few leading companies recently reported upon having started to assess risks for host communities, there is often a lack of detailed and transparent reporting upon the results of such assessments. By 2020, Newmont had initiated a collaboration with the National Center for Atmospheric Research (NSF NCAR) to develop their site-based climate risk assessments. Upon revisiting the more recent reports, we found that Newmont both provided more detail than other companies regarding climate risk at each mining site, including information about the likelihood that the risks would be manifest, and adopted adaptation measures (Newmont 2022). Anglo American highlighted that it had begun to develop a bottom-up approach to complement its tangible risk assessments with the overarching purpose of gaining a better understanding of local climate risks (Anglo American 2022: 19). In 2021, Freeport McMoran (2021: 43) also developed relatively detailed assessments of host communities' climate vulnerability. While these examples suggest that some companies are starting to assess climate risks for host communities, there is little information about how risks are assessed. Overall, our analysis suggests that such climate risk assessments are still in early stages.

One likely reason for the limited progress on integrating RBC in adaptation is staff with expertise on community relations are rarely involved in climate risk assessments, which are typically handled by engineers within environmental units. As stressed by two consultants who were performing community risk assessments for several transnational mining companies, to be able to assess such risks, it is critical to include community relationships units in the discussion:

> The responsibility is typically more on the environmental engineering side than on the community side…We have done work for a client in Ghana, and there they had a powerful realization that environmental and social issues are closely connected. So, where you have a model or a structure within mining companies where those two things come together, our plans work really well (Interview 40).

This quotation and our findings more broadly illustrate that companies generally have not developed internal capacities to assess the societal risks posed by their business activities in the context of climate change.

We have identified meaningful *participation* as another core procedural principle for ensuring that affected communities can influence corporate adaptation responses. Yet, in many cases companies' engagement with local stakeholders is limited to participatory water governance initiatives and has been constrained by communities' limited access to climate risk data. In our analysis of the company reports, we examined various community engagement strategies, including climate risk assessments, participatory water monitoring, and public–private partnerships. We found that only 16.2 percent of the observed companies reported having included relevant stakeholders in at least one of these processes.

The principal form of community participation in adaptation is through various types of participatory water governance initiatives, such as water stewardship and water monitoring. Water stewardship refers to "water [use] that is socially equitable, environmentally sustainable and economically beneficial, achieved through a stakeholder-inclusive process that involves site and catchment-based activities" (The Alliance for Water Stewardship 2022: 4). In one prominent example, South 32 took part in a water stewardship partnership comprised of business, government, and civil society organizations created in response to a severe drought in the uMhlathuze region of South Africa (National Business Initiative 2018). While water stewardship is a promising alternative, it is unclear whether participants in the different initiatives have received relevant and accessible information about how climate impacts exacerbate existing water risks to be able to participate meaningfully in such processes.

Participatory water monitoring is yet another procedure for involving host communities and public authorities in assessing the quality and quantity of water in mining sites. While approaches of this kind can help communities gain a better understanding of the climate-induced water risks associated with business activities, the results are widely seen as manipulated by companies, diminishing communities' trust in the results (Pareja et al. 2019). CSOs and multilateral organizations have consequently been inclined to support more autonomous forms of community monitoring in the mining sector (Interview 42).

In contrast, none of the companies reported upon involving communities in climate risk assessments. This omission denied communities the opportunity to share their perceptions of climate-related risks and thereby influence risk assessments and adaptation measures. In 2021, Freeport McMoran reported upon having started to involve community stakeholders in their North American operations to better understand their concerns about climate-driven risks (Freeport-McMoran 2021: 52). In the same year, Rio Tinto also highlighted the importance of developing procedures for community participation in assessing climate risks (Rio Tinto 2021: 15). These examples suggest that there are emerging initiatives to involve communities in climate risk assessments among a few sustainability frontrunners, but such practices remain isolated and appear to be far from institutionalized across the mining industry.

Finally, as shown in the final bar of Figure 4, which addresses *transparency*, none of the companies in our sample disclosed information about their exposure to climate risks or their adaptation measures. Several company representatives acknowledged the importance of disclosing climate risk data to their host communities, but highlighted the challenges of presenting projections on specific climate-driven risks in a way that is clear and accessible to these groups (Interviews 17, 45). As a company representative observed: "The complexity of climate change, that is a huge barrier for us to be able to have real conversations with communities around the developing world" (Interview 5). In contrast, spokespersons for civil society contend that the prospect of community protests make companies reluctant to share information about how climate change compounds water stress (Interviews 30, 67).

In contrast, 72.9 percent of the companies in our sample disclosed information to investors – through the Task Force on Climate-Related Financial Disclosures (TCFD), the Carbon Disclosure Project, or both – about how climate risks affect their business operations and financial performance. TCFD reports are publicly available on company websites and contain generic information about climate risk assessments and brief examples of risks that have been identified in selected mines. For instance, South 32's report is the most detailed, with 220 climate-related risks identified, yet it lacks information about the nature of the enumerated risks, the mining operations affected, and the adaptation measures taken (South 32 2022: 95). In contrast, Barrick dedicates less than one page to climate adaptation, referring passingly to a few examples of country-level climate impacts (Barrick 2022: 71). Hence, even disclosure initiatives aimed at investors suffer from a lack of transparency and include little relevant information for local stakeholders. Overall there are few indications that affected communities and public authorities have access to information about companies' risk assessments and adaptation measures.

3.3 Conclusion

This section has mapped the extent to which mining companies adhere to the four principles for RBC in adaptation identified in our theoretical framework. It has also contextualized the integration of RBC in adaptation by examining how companies understand and generally respond to climate risks. Using a quantitative analysis of company reports combined with qualitative interviews, we come to three main findings.

First, the majority of the world's thirty-seven largest mining companies see adaptation through the lens of conventional economic risk. Second, about half of these companies have developed some type of organizational response to climate risks. While the most common responses were the integration of climate risks into water governance and risk management, as well as the climate-proofing of infrastructure, few companies considered climate risks in mine closure planning. And, finally, only a minority have integrated the RBC principles into their adaptation responses. Of the ones that did, most companies approached adaptation primarily through conventional CSR investments, rather than incorporating other core procedural elements of RBC, such as risk assessments, stakeholder participation, and transparency. While we see a few emerging examples of sustainability frontrunners beginning to assess climate risks for host communities and to establish limited opportunities for stakeholder participation, none of the companies in our sample were transparent about their climate risk assessments or the adaptation measures they have adopted.

In the following section, we will rely on three in-depth case studies to explain the variation in RBC in adaptation laid out in this section.

4 Explaining Irresponsible Business Conduct in Adaptation

This section examines how the theorized regulatory and political factors have contributed to the persistence of irresponsible adaptation practices, even among sustainability frontrunners in the global mining industry. To this end, we provide an in-depth analysis of three mining projects in Peru: Antamina, Cerro Corona, and Quellaveco. While the first two are operated by companies included in our quantitative dataset, Cerro Corona is run by Gold Fields – a comparatively minor player with a smaller market share. As elaborated in Section 1, our analysis is based on documents and qualitative interviews collected during four weeks of fieldwork in the three mining regions and in the capital, Lima, in 2022.

The remainder of this section is divided into four parts. We begin by providing background information on climate vulnerability and water stress in Peru (Section 4.1). We then lay out the extent to which, and how, companies in the three cases have integrated the RBC principles into their adaptation responses (Section 4.2), which sets the stage for our analysis of the regulatory and political factors that have shaped these outcomes (Section 4.3). Finally, we summarize the analysis against the backdrop of our theoretical expectation (Section 4.4).

4.1 Background: Climate Vulnerability and Water Stress in Peru

To ground our empirical analysis, we briefly describe climate vulnerability and water stress in Peru and in the regions where our three case studies are situated. Landslides, flooding, glacial retreat, droughts, and a wide variety of climate risks have increasingly exposed Peru to climate-related risks, but it is not among the most vulnerable countries in the world (MINAM 2021: 82). According to the Notre Dame Global Adaptation Initiative Index (ND-GAIN), in 2023 Peru ranked 93rd out of 182 observed countries.

Figure 5 shows that climate vulnerability is unevenly distributed across provinces in which the three mining projects are situated, as captured by the PVCCI (for more information on methodology, see FERDI (2022)). The coastal areas, where the Quellaveco project is located, are most vulnerable to climatic risks. However, both Antamina and Cerro Corona are very close to these highly exposed coastal regions.

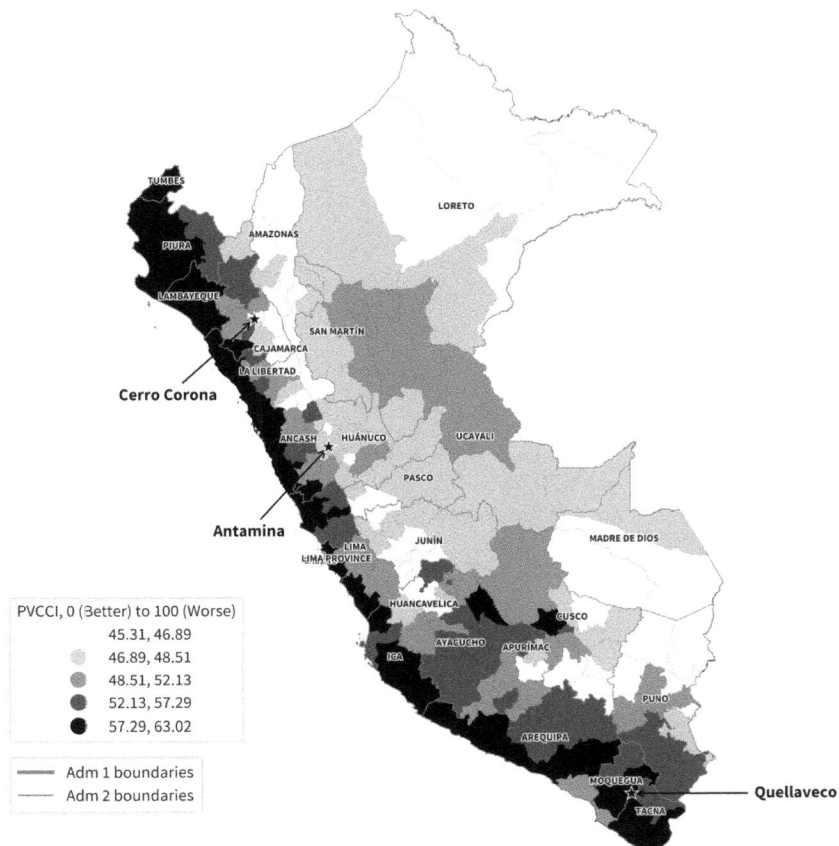

Figure 5 Subnational climate vulnerability in Peru.
Source: Author's mining data.
Note: The map is based on data from the PVCCI (FERDI 2022).

In addition to the risks posed by natural disasters, such as the rainfall extremes caused by an El Niño event, water stress is a pressing concern already being felt in Peru. Figure 6 presents an overview of water stress across provinces that helps illustrate our findings. The map is based on data from the World Resources Institute's (WRI) Water Risk Atlas, which accounts for both surface and groundwater resources (see WRI (2019) for more information on methodology). It shows that water scarcity has already reached an extremely high level in some arid coastal regions and is high across the Andean area. Both Quellaveco and Cerro Corona are located in areas of extremely high water stress, while Antamina operates in an area experiencing medium-to-high water stress.

The coastal regions, where more than half of the population lives and where the country's large-scale export agriculture is concentrated, receive

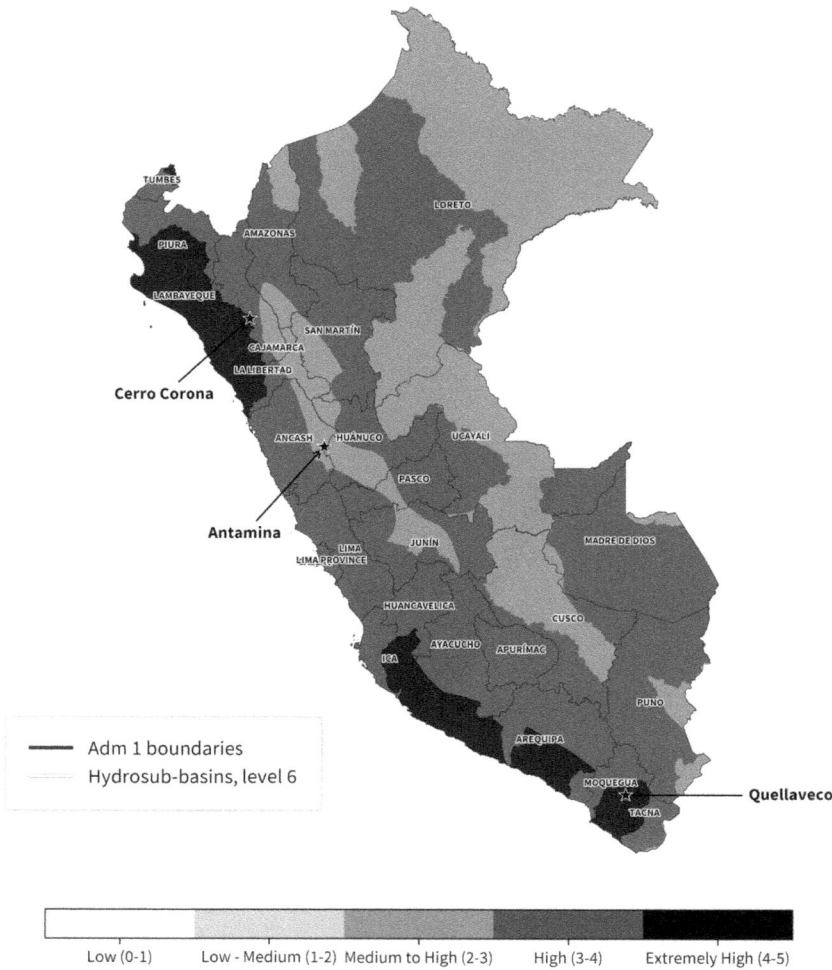

Figure 6 Subnational water stress in Peru.
Source: Author's own mining data.
Note: The map is based on data from the WRI database (WRI 2019).

80 percent of their water from the melting ices of the Andes (Thompson et al. 2021). The Andean area is mainly populated by Indigenous communities and subsistence farmers known to be highly vulnerable. Increasing precipitation variability has made rain-fed agriculture more difficult, leading to increased demands for irrigation systems and water reservoirs both in the coastal areas and the Andes (Bury et al. 2013).

Since the early 2000s, Peru has sustained more severe climate impacts along with a rapid expansion of mineral extraction in the Andean highlands. Peru is currently the world's second largest producer of copper, silver, and zinc; and it is also a leading producer of gold, lead, and molybdenum in Latin America (MINEM 2022). Mining has thus been a key factor in reducing poverty and sustaining economic growth and has placed Peru among upper middle-income countries (Gustafsson 2018).

At the same time, mining activities have placed immense pressure on water supplies and livelihood assets, incited social conflicts, and contaminated the environment in affected areas (e.g., Bebbington and Bury 2009). For example, Peru's Ministry of the Environment (MINAM) has identified 7,668 environmental liabilities, which are toxic waste deposits left unremediated in the aftermath of mine closures (MINAM 2021). During periods of heavy rain, toxic substances risk contaminating the surrounding land and water, thereby further exacerbating the climate vulnerability of local communities. How mining companies respond to these different climate-related impacts is thus likely to be intrinsically linked to the climate resilience of Andean communities, coastal export agriculture, and urban areas. Such risks underline the fundamental importance of integrating the RBC principles into their adaptation responses and underscore that Peru is a particularly important case for examining our theoretical framework. Focusing on Peru helps understand how public governance, voluntary standards, and the nature of civil society advocacy have motivated sustainability frontrunners to compensate communities for climate-related harms through conventional CSR, rather than by integrating procedural principles.

4.2 Comparing Corporate Adaptation Across Three Subnational Contexts

Next, we introduce our three cases and analyze to what extent and how companies have integrated the theorized procedural and distributive RBC principles into their adaptation responses. We show how these responses are shaped by the often conflicting dynamics of extractive activities within the distinct subnational contexts the companies operate in. This analysis sets the stage for the empirical analysis of the explanatory part of our theoretical framework.

Antamina: A Reactive Approach to Local Climate Concerns

Antamina is operated by a Peruvian mining company owned by Glencore, BHP, Teck, and Mitsubishi. As we show in Section 3, these companies, most notably Teck, belong to the small group of industry actors that have begun to integrate the RBC principles into their adaptation responses. In Peru, mining experts recognize Antamina as a company that employs dialogues, partnerships, and social investments as a general approach to address community concerns. One expert on community rights in the mining sector told us that this approach was an important explanation for the absence of major protests and strong anti-mining mobilizations against the company over the years (Interview 57). While this general approach to local communities is to some extent reflected in its adaptation responses, the procedural principles remain weakly integrated.

Antamina extracts copper and zinc from its mine in the Ancash region, located at an altitude of 4,300 meters above sea level. The extracted minerals are then transported via pipeline to the shipping port in the coastal province of Huarmey, which has led to some concerns among communities affected by the pipeline. Ancash is the region with the largest area of tropical glaciers in the world. Due to climate change, 35 percent of the total glacier-covered area has been lost over the past forty years with severe consequences. For the human population, it has significantly reduced the availability of freshwater during the dry season with manifold repercussions (Gobierno Regional de Ancash 2017: 7). Mining activities have further exacerbated water insecurity for downstream populations (Bury et al. 2013). Intense rains resulting in landslides pose another salient climate risk (Gobierno Regional de Ancash 2017: 7). The most noteworthy climate-related event occurred in 1970, when a devastating earthquake caused the death of about 80,000 people. The history of environmental disasters, melting glaciers, and the presence of NGOs working on adaptation projects have raised awareness of climate impacts and natural disasters throughout the local population. Demonstrating its responsible management of such climate-related risks was, therefore, essential for Antamina to gain social license to operate.

Our analysis of how the company has handled climate-related risks reveals that Antamina (2020) has not reported carrying out climate risk assessments for its own operations or *assessing climate risks to society*. Compared to other companies that combine global climate-driven scenarios with site-level data to analyze the likelihood and prospective consequences of climatic impacts, Antamina does not appear to be working with climate data suited for making predictions. This makes it hard for the company to proactively identify the risks for host communities linked to their operations. At the same time, the company does refer to climate impacts in its assessments of water and disaster risks. Instead of anticipating future conditions, it concentrates on climate-related impacts that have already begun to affect local communities within its area of operation (Antamina 2020), illustrating its reactive approach to such risks.

Antamina has, to some extent, integrated the procedural principles by being *transparent* and *participatory* in specific adaptation responses. Community members have been concerned about the risks of accidents related to mining infrastructure. In 2019, there were indications that the tailings dam could collapse, and the entire community had to be temporarily evacuated. The dam, which contains highly toxic topic mining waste, is located just above the community, and its collapse would have had catastrophic effects. That incident was followed by dialogues between the company and the community. These dialogues led to an agreement that Antamina would both improve dam stability to be able to withstand more extreme weather events and develop a Community Early Warning System to alert members of the local population of an emergency. Another example of a CSR investment is the climate-proofing of the pipeline that transports minerals to the coast along a route that passes through an area highly exposed to climate-related disasters (Cenepred 2023). In 2019, intense rains damaged the pipeline, which led to community protests about risks of contamination of land and water (CooperAcción 2021). In response, the company developed a partnership with public institutions working on disaster risks and conducted a risk assessment. These actors reached an agreement that the company would climate-proof the pipeline to reduce communities' exposure to the risk of dangerous accidents (Interviews 62, 73, 74).

In both cases, Antamina provided local communities with information and engaged them in the planning and implementation of adaptation measures. Antamina informed the community about the improvements it made to the tailings dam, involved community members in the development of an Early Warning System, and organized a public consultation about plans to climate-proof the pipeline (Interview 73). The company has also set up participatory water monitoring to enable communities to regularly assess the operation's impact on water quantity and quality. These examples suggest that Antamina has, to some extent, integrated the procedural principles by being *transparent* and *participatory* in certain adaptation responses. However, such steps were only taken after repeated community demands on the company to disclose data and engage in dialogue.

Importantly, community members lacked trust in the information the company provided. When it reassured them that the upgrade of the tailings dam was sufficient, the community engaged an external technical advisor who maintained that the company had underestimated the risks to such an extent that the entire community should be relocated (Interview 76).

Similarly, one community leader told us that they were suspicious of the accuracy of company results, as the company water assessments always showed that the level of contamination never exceeded the legal limit (Interview 63). Most importantly, these examples reveal that Antamina has not proactively consulted local communities about climate risk assessments or adaptation measures. Instead, it has disclosed information only under pressure – and even then, limited disclosures to the most visible climate impacts, particularly those too evident to ignore or deny.

Finally, Antamina has made *CSR investments* aimed at compensating local communities and strengthening their resilience in response to climate impacts that have become a growing source of local concern. For example, the company has installed water reservoirs for communities near the mine, and in areas where the pipeline passes, Antamina has partnered with the Ministry of Agriculture and local governments to implement a project aimed at improving water governance and irrigation infrastructure and raising awareness of climate change.

Overall, Antamina has invested in public adaptation goods through CSR initiatives, but its integration of key procedural principles has been limited and often reactive – typically occurring in response to the threat of community protests.

Cerro Corona: Corporate Adaptation in Contested Terrain

Gold Fields, in contrast to Antamina, adopted a more proactive and comprehensive strategy to integrate our four RBC principles into its adaptation strategy for the Cerro Corona project. The company emphasizes community resilience as a cornerstone of its adaptation policy (Gold Fields 2021a: 11). Interviewees both within and without the company have remarked that the CEO and the sustainability manager for the Cerro Corona project displayed a high level of commitment to implementing the policy (Interviews 6, 40, 36, 37, 59, 60, 78). That commitment notwithstanding, the company largely abandoned an important component of its work on community adaptation due to unfavorable sociopolitical conditions and staff changes in the associated CSR unit.

The Cerro Corona gold and copper project is located in the impoverished mining region of Cajamarca, where rural protests over mining impacts on scarce water resources have led to the indefinite postponement of the development of two large-scale mines. There have, to date, been few protests against the Cerro Corona project. According to the regional climate change plan, climate vulnerability is classified as "high" or "very high" in 85 percent of the region's territory, with droughts, floods, and frosts identified as the most significant risks (Gobierno Regional de Cajamarca 2016: 35). Heavy rains resulting in landslides are also a key risk that must be accounted for in any assessment undertaken (Gobierno Regional de Cajamarca 2016: 43). The Cerro Verde project is located in Hualgayoc, the district with the highest number of environmental liabilities in Peru. It is the site of a staggering 1,036 unremediated environmental liabilities, which enable toxic substances to pollute the surrounding water and land during periods of intense rainfall. This situation led MINAM to declare the district an area of environmental emergency in 2016 (MINAM 2016). More broadly, water stress is becoming an acute problem in both rural and urban areas. In the urban area of Hualgayoc, the population relies entirely on Gold Fields' provision of freshwater in tanks, and within rural communities the key demand is that Gold Fields invest in water reservoirs (Interview 54).

Gold Fields has committed to supporting community resilience by integrating all RBC principles. Since 2016, Gold Fields has conducted two *assessments of climate risks* comprising both their own operations and those of surrounding communities, which led them to adopt several adaptation measures (Gold Fields 2022: 77). To carry out the site-based risk assessments, the company piloted the MiCA tool developed by ICMM, which enables companies to analyze climate risks, including community vulnerability; and it also engaged external consultants to conduct community climate risk assessments (Interview 40). Gold Fields reported having addressed the identified risks by reducing its water consumption and improving tailings management in the Cerro Corona project (Gold Fields 2021b: 83). Compared to other companies in our sample, the company has a more advanced understanding of community resilience, particularly in its approach to tailings management.

Cerro Corona focused most intensively on integrating the RBC principles between 2014 and 2017, when it implemented the "Adapting Together" program in partnership with NGOs and the local government. The project involved building institutional capacity, developing an adaptation plan, and formulating adaptation projects eligible for funding via the tax revenues transferred to mining-producing subnational governments. It also constructed 250 water reservoirs to support agricultural activities, a potable water system for the city of Hualgayoc, and developed training programs to increase the awareness of climate risks within local communities (Goldfields 2019: 112; Lutheran World Relief 2017). This shows how the company used *CSR investments* to support adaptation infrastructure, awareness-raising, and capacity building of subnational governments.

Importantly, this initiative was implemented as an institutionalized partnership, rather than being targeted at smaller groups critical of the mine.

The initiative also included components that may support the integration of the procedural principles, such as *transparency* and *participation*. Gold Fields distinguishes itself among the three companies studied by making technical information accessible and engaging in dialogues about climate change. For instance, it used local radio, theater, and brochures to disseminate information about climate impacts and adaptation to local communities. A company representative explained that the goal was to find "a neutral platform with a collaborative space where we can share our technical data on [climate] risks so that the farmers understand that there is something bigger […] something that we are also battling" (Interview 36). A former sustainability manager at Cerro Corona described it as an attempt to create a win-win situation where the mine is "viewed as an opportunity to adapt to climate change rather than a risk" (Interview 78). However, this framing of shared risks was met with suspicion by local politicians and communities (ibid). One community leader who participated in the training saw it as an attempt to shift blame for the impacts caused by mining to climate change (Interview 53), suggesting that such CSR initiatives did not have the intended impacts of increasing community support for the mining project. Local communities, though aware of longer dry periods and more intense rains than in the past, voiced far greater concern over the deeply unsustainable effects of mining in their regions, which they perceived as the main cause of the climate-related problems they now faced.

The deep-rooted distrust against the company was an important barrier to integrating the procedural principles. Gold Fields claims to be regularly disclosing information about the mine's impacts on water quantity and quality and involving local communities and authorities in participatory monitoring (Gold Fields 2019: 106). But community members are skeptical about the information the company releases on water impacts (Interviews 54, 56). A representative from the regional government admitted having never seen a test result indicating that the company was tied to any polluted water detected, raising suspicions that the results had been manipulated (Interview 70). Moreover, the company failed to invite community members or representatives of public authorities to participate in climate risk assessments (Interviews 36, 40). Yet, in line with the newly adopted Global Industry Standard on Tailings Management, Gold Fields has publicly committed to involving local communities in climate risk assessments of its tailings dams (Gold Fields 2021c: 15). However, whether this has actually been done has yet to be determined.

Overall, the Cerro Corona case illustrates an attempt to integrate the four principles. At the same time, this case also reveals the challenges associated with disclosing information and engaging in dialogues about climate adaptation in contexts where past and present mining activities are deeply implicated in societal climate vulnerability. In such situations, companies are likely to be unwilling to disclose information about the risks associated with mining activities in a context of climate, as it might lead to community protests.

Quellaveco – Adapting Core Business Activities

Quellaveco, operated by Anglo American, had recently committed to integrating the RBC principles into its adaptation strategies during the time we conducted our fieldwork in 2022. Although Anglo American describes this work as evolving, it claimed it was committed to integrating climate adaptation in its social management systems (Anglo American 2022: 20). Its approach differed from Cerro Corona's in that the adaptation measures were more closely linked to the company's core operations, partly because they were already integrated into the project's initial design. The company made relatively significant CSR investments in adaptation, but generally fell short in the integration of procedural principles.

The Quellaveco copper project is located in the Moquegua region and currently constitutes the largest mining investment in Peru (MINEM 2022: 12). Consistent with its scale, mining plays a crucial role in the regional economy and, as of 2018 (the most recent year for which data is available), 87 percent of that region's revenues originated from mining (Arellano-Yanguas 2018: 5). Although the population generally recognizes the importance of mining as a driver of regional development, concerns about water stress continue to be a pressing issue. Moquegua is located in the Atacama Desert, where water scarcity is extremely high (see Figure 6). Despite these challenging conditions, extending the agricultural frontier into the arid desert remains a regional priority. Following a lengthy negotiation process, during which the company made significant concessions to reduce the project's impact on the region's water resources and made CSR investments to facilitate the Pasto Grande irrigation project, the Quellaveco project finally began operating in 2023 (Gustafsson and Scurrah 2023).

Discussing the first procedural principle, Anglo American began *assessing climate risks* for both their own operations and for its host communities in 2010. The results of this assessment showed that 70 percent of

Anglo American's mines were located in water-scarce areas, with the highest risks to their projects in the Atacama Desert in Chile and Peru (Anglo American 2018: 49). Consequently, the company's adaptation strategy has focused on reducing its water footprint (Anglo American 2022: 9). A company representative in the environmental unit explained that Anglo American made a generic climate adaptation study as part of the EIA it conducted based on data from public institutions, but planned to undertake a new climate risk assessment in 2023 based on their own local data. To this point the company has not involved local communities in its climate risk assessments (Interview 85).

To address concerns from local communities and the regional government about the potentially negative impacts of the Quellaveco mine, Anglo American made significant concessions by modifying the project's engineering design. In the initial EIA presented in 2000, the company proposed using groundwater for its operations. But the results of an independent report by an international water specialist commissioned by the regional governor showed that the project would produce a severe water shortage. In 2008 Anglo American presented a new project design that would rely on water from the Titire River – a waterway naturally contaminated by boron and arsenic and therefore not suitable for consumption – rather than groundwater.

Between 2011 and 2012, the company participated in a sixteen-month-long dialogue table with members of the regional government and civil society representatives. The process resulted in twenty-six agreements and the establishment of a committee dedicated to monitoring company compliance (Jaskoski 2014). An important concession that emerged from these talks was Anglo American's commitment to build an eight-kilometer-long tunnel to direct the Asana River around the mine. The dialogue table was led by the regional government and was carried out in a relatively *participatory* way, resulting in a final agreement that reflected a broad regional consensus.

In addition to substantial changes in the project's engineering design, Anglo American also committed to making significant *CSR investments* in water infrastructure which, according to company representatives, will positively affect the water balance in the area (Interview 85). To accomplish this, the company has constructed the large Vizcachas large dam to store rainwater (Anglo American 2018: 48). The dam was critical for the acceptance of the project since the water could be used for an irrigation project to enable the extension of Moquegua's agricultural frontier. However, there are also concerns about the long-term sustainability of the dam with rainfall steadily decreasing and becoming increasingly irregular

(Interview 57). In 2021, there was not enough rainwater to fill the dam, which led to water shortage (Interviews 79, 83).

Another significant *CSR investment* was made in the development of a technical solution that would enable the contaminated water from the Titire river to be treated and clean water returned for use by farmers in the downstream Tambo Valley. Anglo American claims that its approach, once implemented, will improve the water quality for the Tambo region. However, farmers who would be affected fear that the cumulative volume of river water would be reduced and consequently have protested against the project and appealed the decision of the National Water Authority to grant the company a water license (CooperAcción 2022). A former leader of a rural organization in the Tambo Valley claims that the decision to issue the water license was based on 2009 data that is no longer reliable and does not consider the impacts of climate change on hydrological stress (Interview 75). CSOs rarely express such an open critique of the lack of regulatory controls on companies' adaptation strategies, which also eases the pressure on public authorities to adopt stricter controls on business activities to reduce climate-related risks.

It is noteworthy that Anglo American has negotiated in a relatively *participatory* and inclusive way, made important concessions, and dedicated significant *CSR investments* to water infrastructure. Yet it is equally notable that it has not been *transparent* about the climate-related risks linked to their operations. The company has not disclosed the results of its climate risk assessments to relevant public authorities or local communities, leaving them without the information necessary to evaluate the risks associated with mining activities in a context of climate change. Even though water stress is a key climate risk and the project has been designed to reduce its impacts on water resources, climate adaptation was not discussed explicitly with local stakeholders. Two civil society representatives we interviewed were concerned about the lack of a public debate on the long-term sustainability of the company's adaptation strategies. They warned that the climate-related risks associated with the project had been significantly underestimated. They were also worried that water stress would severely affect local populations in the future (Interviews 47, 81). This suggests that the lack of *transparency* regarding the short- and long-term effects on water stress is an important shortcoming in Anglo American's approach.

Overall, while Anglo American has made significant CSR investments in adaptation and has taken some steps to assess climate risks to society and engage in participatory process, it has not been transparent about the risks associated with mining in a context of climate change.

Table 1 Companies' adherence to the principles of responsible business conduct.

	Antamina	**Cerro Corona**	**Quellaveco**
Assessments of climate risks to society	Partially	To large extent	Partially
CSR investments	Significantly	To large extent	To large extent
Participation	Partially	Partially	Partially
Transparency	To little extent	To little extent	Not at all

Table 1 presents an overview of the integration of the principles in the three cases. The findings are consistent with our analysis in Section 3. The *assessment of climate risks to society* is a relatively new practice in each of the three cases. Compared to the other companies, Cerro Corona followed a more systematic approach in seeking to include community vulnerability in its climate risk assessments. By contrast, Antamina had a more reactive approach, while Quellaveco offered only a generic analysis of community vulnerability, which inevitably limits its utility as a tool for proactively identifying climate-related risks. CSR investments in public adaptation goods appear to be the most systematically integrated principles in all three cases. Finally, although each company has participatory procedures for involving local communities in addressing some climate-related risks, specifically those relating to water, they have disclosed only selective information about climate risks and adaptation interventions. Without access to such information, it is difficult, if not impossible, for local communities or governmental agencies to hold companies accountable for how they handle the risks linked to mining activities in a context of climate change.

Taken together, the findings suggest that there are both differences and similarities in RBC in adaptation between the three companies. All have primarily addressed climate risks through conventional CSR strategies rather than through procedural components of RBC, such as risk assessments, stakeholder participation, and transparency.

4.3 How Complex Societal Risks Become Business as Usual

Next, we seek to explain the outcome analyzed earlier, namely, that companies have focused on CSR investments, but have rarely been transparent about climate risks associated with their business activities. We analyze how each of the three explanatory factors – public governance, voluntary adaptation standards, and civil society advocacy – has shaped RBC in adaptation by comparing them across the three cases, both separately and in combination.

Public Governance

Climate governance in Peru is in its early stages. Although the country has taken a progressive stance in international climate negotiations, progress in the domestic policy arena has been limited. In 2014, Peru hosted the twentieth session of the Conference of the Parties (COP20) and played a central role in the multilateral negotiations that led to the Paris Agreement in 2015 (Castro Pereira 2023).

In 2018, Peru passed the Climate Change Framework Law (Framework Law No. 30754) – its first framework law on climate change – which established a relatively ambitious structure for public policies on climate adaptation and mitigation. For its work on climate mitigation, Peru has received substantial international funding for forest conservation in the Amazon, and climate mitigating has been a key priority for MINAM (Interviews 16, 48, 69). However, there is an urgent need to accelerate adaptation efforts given the country's climate vulnerability and the severe water stress (Figures 5 and 6). In 2021, the National Adaptation plan was adopted (RM 096-2021). MINAM's efforts to implement the adaptation plan have, however, been undermined by a lack of resources, a paucity of data on climate impacts, the weak institutional capacities of regional governments, and the counter-lobbying in which economic and political elites have engaged (Interviews 50, 58). Against this backdrop, the Peruvian government has sought to steer corporate adaptation efforts via multistakeholder initiatives and dialogues rather than regulatory controls.

The Ministry of Energy and Mining (MINEM), which is responsible for regulating large-scale mining, has so far done little to incorporate climate adaptation into sectoral governance instruments, leaving it up to companies to voluntarily integrate the RBC principles in their adaptation responses. For example, in EIAs the anticipated impacts of a mining project are evaluated, and companies are bound to disclose the results of risk assessments and organize public consultations. However, the Peruvian EIA regulation only requires companies to present a climate change study, while – in contrast to other types of risks – there are no requirements to establish a management plan to redress the climate risks identified (Law no. 27446). The National Environmental Certification Service for Sustainable Investments has developed voluntary guidelines for integrating climate adaptation measures into EIAs. However, according to two interviewees working at the state agency that monitors the mining sector, without enforceable legal obligations, the uptake of guidelines like these has been limited (Interviews 70, 71). Similarly, there are no legal requirements that climate impacts be considered, be transparent, and enable stakeholder participation in the design of tailings dams and closure plans (Interview 69, 72). The National Water Authority does not take future climate impacts on water resources into account when approving water concessions for mining companies, which implies that the state has no information about intersecting impacts between climate change and mining activities (Interview 32). Consequently, there is no instance of the Peruvian government limiting or prohibiting mining activities that could have adverse impacts on society or the environment.

Indeed, representatives of the companies we studied all agreed that neither existing nor the threat of future state regulations have influenced their adaptation strategies; under the circumstances, they were not perceived as providing any meaningful incentives for them to integrate RBC in adaptation. Both Antamina and Gold Fields have occasionally disclosed climate risk assessments to local communities, but none of the companies have institutionalized such procedures, as previously discussed. Further diminishing its utility, the information that is disclosed tends to be fragmented and selective. As a representative at Antamina argued: "As it's not a legal requirement, not even in the EIA, it's just a question of voluntary social responsibility" (Interview 80, see also 85). A representative of Gold Fields both underscored the need for state regulations to ensure that companies address climate risks in a transparent and inclusionary way, but recognized that the majority of mining companies were strongly opposed to such regulations (Interview 78). Our findings suggest that the absence of strict legal requirements has been, and continues to be, a critical impediment to fostering RBC in adaptation.

Instead, MINAM and regional governments have sought to incentivize companies to engage in climate adaptation through dialogues and partnerships. While such processes could potentially foster greater transparency and participation, national and subnational officials emphasize that these initiatives are mainly used as a means of persuading large companies, specifically in the mining sector, to commit financial resources toward the implementation of adaptation plans, particularly by investing in water storage infrastructure (Interviews 59, 64, 69). According to an interviewee responsible for climate adaptation in Cajamarca, subnational government officials were inspired by Gold Fields' adaptation project and considered inviting mining companies to help implement their own climate adaptation plans. However, others stressed the risks of business capture of the adaptation agenda, which could delay or even derail discussions on the ways mining activities might undermine societal resilience (Interviews 48, 49).

To promote greater private sector involvement, national authorities have relied on soft steering and established multistakeholder initiatives and partnerships. In 2021 MINAM formed a steering group aimed at engaging the private sector in implementing the Nationally Determined Contributions, which are plans laying out national strategies for climate action under the Paris Agreement. In the steering group, twenty business associations participate regularly (MINAM n.d.). The Ministry of Agriculture, through the Sierra Azul climate adaptation fund, has promoted public–private partnerships, notably by granting tax deductions for contributions to water storage infrastructure for small farmers. Antamina is among the mines that benefitted from partnerships of this type (Antamina 2017). With water infrastructure being a key demand of local populations, representatives from Anglo American and Gold Fields have also shown interest in participating in a climate-related partnership (Interviews 78, 85). Public institutions have used this as a means to incentivize companies to make CSR investments in public adaptation goods. Earmarking funds to this end aligns with companies' need to legitimize their presence by showcasing their positive impacts on local development.

Another important soft steering mechanism is the provision of scientific data on climate risks. Public institutions have used existing public databases on disaster risks to support companies to assess climate risks to society and thereby reduce the risks of dangerous accidents. The National Center for Estimation, Prevention and Reduction of Disaster Risks, which delivers high-quality scientific data on regional climate disaster risks, partnered with Antamina to develop a Community Early Warning System and supported the company in climate-proofing the pipeline it uses to transport minerals. The Environmental Assessment and Enforcement Agency of Peru has started to collect data on extreme weather events to alert companies about emergencies that could lead to the collapse of tailings dams (Interview 70). These examples show how public institutions have sought to steer companies to integrate the RBC principles by providing scientific information about climate risks.

However, in the absence of any legal requirements, the use of such data is likely to be limited. Moreover, while there is high-quality data on climate disasters, the National Water Authority has no information on the effects of climate change on water stress. This lack of data is a significant limitation, as this is the most pressing climate impact in many mining localities in Peru.

Although the three companies interact closely with regional and local governments, company representatives did not perceive climate adaptation as a salient topic in any of the subnational regions (Interviews 78, 80, 85), which is likely because the three regions are still in the process of developing their adaptation plans. In the case of Cerro Corona, company representatives highlighted a lack of support from the local government as the main reason they stopped working on climate adaptation governance (Interviews 60, 78). None of the subnational governments have used formal adaptation planning institutions as a way to steer companies' adaptation strategies. Moquegua stands out among the regional governments in having negotiated an agreement in which Anglo American committed to making major investments in water infrastructure and changes in the engineering design of the project. By conditioning the acceptance of the project on these changes, the regional government contributed to improve the company's adaptation strategy (Gustafsson and Scurrah 2023).

Overall, public governance imposes no legal restrictions on companies in the area of adaptation. Instead, it primarily incentivizes the use of CSR investments through partnerships and, to a lesser extent, through the provision of scientific data about climate-induced disasters. However, there are also important differences among the cases. In the case of Quellaveco, the regional government played a significant role in shaping how the company addressed climate-induced water stress. This influence can be attributed to the regional government's capacity to generate and deploy scientific data to evaluate the project's impact under conditions of climate-induced water stress, which gave it greater leverage in negotiating CSR investments in water infrastructure with the company. In the other two cases, subnational governments had less influence on corporate adaptation strategies. They mainly entered into partnerships with the companies to incentivize them to make CSR investments in water infrastructure. Taken together, and in line with our expectations, we found significant barriers related to the absence of regulatory controls and the limited capacity of national authorities and subnational governments to use soft steering to influence corporate adaptation behavior, except for making CSR investments.

Voluntary Adaptation Standards

As per our theoretical expectation, without legally binding rules, emergent voluntary adaptation standards can contribute to the incorporation of the RBC principles in companies' risk management systems through monitoring of compliance and capacity-building incentives. While the quality of compliance monitoring varies widely, at the very least standards can contribute to the spread of norms and best practice about responsible business behavior in adaptation (Marx et al. 2024). In the context of adaptation, standards have primarily appeared in response to investors' concerns about the effects of climate impacts on operational resilience and the sustainability of business practices, suggesting that market incentives have generated a demand for such standards. For instance, the Equator Principles, supported by national and international financial institutions, require companies to consider climate risks in EIAs for all major projects (Equator Principles 2020). With climate impacts identified as one of the most serious threats to business resilience, investors are increasingly in search of reliable data on how companies handle climate adaptation. A variety of private voluntary standards have emerged to support companies in meeting investor requirements. For instance, in 2015, the Financial Stability Board established the TCFD, which was widely used by companies until it was disbanded in 2023. It was replaced by the International Financial Reporting Standards, which will further develop climate-related disclosure. In the mining sector, ICMM has played an important role in developing industry-specific standards for adapting to climate-induced change and provided a valuable platform for exchanging experiences in the larger arena of climate adaptation.

Representatives of both Gold Fields and Anglo American emphasized the critical capacity-building role of ICMM's standard and underlying investor pressure in the development of their adaptation strategies (Interviews 78, 85). A representative of ICMM recognized Gold Fields as a company that had been particularly active in developing and implementing ICMM frameworks on climate adaptation and community resilience (Interview 6). Likewise, an interviewee speaking on behalf of Antamina recognized that ICMM frameworks had been an important reference point for their work on adaptation (Interview 80).

Whereas private voluntary standards have clearly influenced the adaptation strategies of the three companies, standards differ in the extent to which they integrate RBC principles. For instance, reporting in line with the TCFD is aimed at increasing transparency vis-á-vis investors, yet the information disclosed is too generic to provide the affected parties and governmental agencies much, if any, relevant information about site-specific climate risk assessments and adaptation strategies. One contributing factor is likely to be the absence of a clear definition of "adaptation," which has given companies significant discretion in presenting their findings on risk exposure and adaptation measures. In turn, differences in the level of detail incorporated into their adaptation reports are readily discernible. For example, Gold Fields provides a detailed account of identified risks, adaptation interventions, and the progress achieved (Gold Fields 2022). In contrast, Anglo American presents only general country-level risks on a purpose-specific map, without offering any information on prioritized risks or adaptation interventions (Anglo American 2022). The same applies to the owners of Antamina (Glencore 2021; Teck 2022). This suggests that TCFD has not contributed to provide local communities and governmental agencies with relevant and accessible information about climate risks and adaptation measures. Hence, existing voluntary adaptation standards have done little to improve transparency.

Compared to the TCFD, ICMM provides more detailed specifications on RBC principles in its standard. However, its requirements remain too lax, lacking the concrete indicators and monitoring mechanisms to significantly improve corporate adaptation. For example, according to ICMM's principles, companies should "engage with host communities on our shared climate change risks and opportunities and help host communities to understand how they can adapt to the physical impact of climate change" (ICMM 2021: 5). By 2016, ICMM had launched a Climate Assessment Data Tool that companies could use to assess the climate risk exposure of their own operations and for that of surrounding communities (ICMM 2019). This could potentially be used as a means both to disclose information and to involve local communities in assessing climate risks and planning for adaptation. However, use of this tool is entirely voluntary, with no specific requirements or indicators guiding companies on how to engage host communities in participatory adaptation processes. As a result, while Gold Fields piloted the tool, it did not involve affected communities in its climate risk assessments. Ultimately, the lack of specific indicators and strong enforcement mechanisms limits the effectiveness of these standards in shaping corporate behavior.

By contrast, water stewardship and tailings dam standards are included among key performance indicators, with company compliance monitored by independent auditors. These areas have relatively detailed requirements, including on how to engage communities in planning processes (ICMM 2020: 27). In our three cases, company representatives emphasized that they had made some changes to comply with the new tailing dam standards, suggesting that the enforcement mechanism has generated some level of compliance (Interviews 78, 80, 85).

In contrast, compliance with water stewardship and other adaptation-related aspects of the ICMM standard appears to be subject to interpretation and internal prioritization. Gold Fields representatives emphasize that their decision to focus on climate adaptation and community resilience was made early and taken seriously by key decision-makers (Interviews 36, 37, 78). Similarly, representatives of Anglo American affirm that climate adaptation is a priority, while acknowledging that their primary focus has been on addressing the urgent issue of water stress (Interview 85).

In the end, ICMM appears to prioritize capacity-building in adaptation while allowing companies significant flexibility in interpreting and constructing compliance with the standard. It seems to be intended to influence company behavior by learning and support, rather than imposing burdens and controls on companies. As a former ICMM employee remarked, "there is always a tension between minimizing the costs for the companies and improving sustainability within ICMM [...] and they are always more inclined to listen to the demands of the investors" (Interview 87, see also interview 78). The absence of clear and measurable indicators of adaptation has made meaningful compliance dependent upon the efforts of internal champions to convince other firm actors to comply. Norm entrepreneurs of this kind were especially active players in the Cerro Corona case until 2020, when a community relationship team less committed to community resilience took over (Interview 78).

Compared to the OECD guidelines that stipulate that companies have an obligation to avoid negative socio-environmental impacts when adapting to climate change (OECD 2023: 37), ICMM's adaptation standard is more lax. ICMM portrays climate impacts as a shared risk and presents the business case for voluntarily supporting community adaptation. This framing allows for the decoupling of climate impacts from business activities, positioning companies as affected groups rather than as being responsible for exacerbating such risks. Generally, ICMM favors a voluntary CSR approach to climate adaptation rather than one centering on an obligation to do no harm.

To conclude, in the absence of domestic legislation, private voluntary adaptation standards began to emerge after 2015. These standards have played a key role in spreading norms and promoting learning about RBC in adaptation, particularly in the case of Gold Fields. However, adaptation is only partially included among audited performance indicators, and weak enforcement mechanisms grant companies significant discretion in interpreting ICMM's standards. As a result, meaningful compliance often depends on the commitment of key decision-makers within these companies. In all, our findings in all three cases support our expectation that companies will rely on voluntary standards to incorporate the RBC principles, but our analysis suggests that their influence is weak as the majority of existing standards are not accompanied by strong enforcement mechanisms.

Civil Society Advocacy

Although voluntary adaptation standards have spread awareness about the RBC principles, there has been a noticeable absence of civil society demands for companies to align their actions with these principles. Instead, CSO advocacy has mainly centered on companies' negative impacts on water resources, leading companies to offer CSR investments in water infrastructure as compensation for negative impacts. In contrast, CSOs have rarely pressured companies either to enable stakeholder participation or to become more transparent about their adaptation-related actions. In the three cases we examined, the relative strength and capacity of CSOs to effectively negotiate with companies differed.

Although the region where Cerro Corona is situated has strong movements that have blocked several large-scale mines, the project has not been the movement's focus. In addition to being localized and somewhat fragmented, CSO attempts to negotiate with the company have tended to prioritize CSR investments. Local communities are divided between those that have benefited from CSR investments and the labor opportunities they provide and those that have not. Their principal demand has been to construct new water reservoirs or improve existing ones (Interviews 54, 55). Recognizing that water scarcity is a very sensitive issue, representatives of Gold Fields have focused on reservoir-building (Goldfields 2019: 112; Interview 78). Communities have also raised concerns about water pollution. For instance, the river pollution that ensued after the leakage of the tailings dam in 2018 prompted communities to ally with the municipal government and rural organizations in the region to file a legal complaint against the company, which had not been settled as of late 2024 (Interview 56). In this and other cases, communities perceived climate impacts as inseparable from the existing water-related impacts of mining activities rather than as a qualitatively new phenomenon.

Likewise, there has been no large-scale mobilization against the Antamina project. Sporadic community protests have focused on specific mining-related impacts that have been exacerbated by climate change. It was only after communities raised concerns about the tailings dam that the company developed an emergency plan. Protests related to risks associated with the pipeline led the company to take appropriate climate-proofing measures and invest in several adaptation projects. Company representatives we spoke to recognized that concerns about community protests were critical to Antamina's decision to remediate these climate impacts (Interviews 61, 62, 80). Here as well, localized protests played a significant role in the company's decision to disclose pertinent information, involve communities in adaptation planning, and invest in CSR initiatives meant to improve water infrastructure.

In the case of Quellaveco, CSOs were able to form a broad-based coalition with local elites, business groups, NGOs, rural organizations, and the regional government that demanded the company to find a solution to reduce its impacts on scarce water resources. In response, the company changed the project's design and invested in the water harvesting infrastructure needed to extend the agricultural frontier in the Atacama desert (Anglo American n.d.).

Our findings suggest that civil society advocacy has often foregrounded the impact of mining activities on water resources and lead to strong demands for CSR investments in water infrastructure. Climate change and adaptation have largely been absent from CSO demands, which helps explain why there has been so little progress toward institutionalizing the other procedurally oriented principles. Although CSOs recognize the impacts of climate change on water stress, they emphasize that mining companies are principally responsible for the current water scarcity (Interviews 56, 63, 67). In each case, local communities have concentrated their efforts on providing scientific evidence showing the impact of mining activities on water quantity and quality, challenging companies to become more transparent about their negative impacts on water. In the presence of competing expert knowledge, it has proven difficult to demonstrate causality for negative impacts, as, for instance, illustrated in the case of the pollution for which Gold Fields was responsible (Interviews 53, 63, 79).

In contrast, CSOs have rarely demanded that companies become more transparent and inclusionary in their responses to climate impacts. Mobilizing around issues where it is difficult to attribute responsibility to specific actors is generally more challenging (cf. Carpenter 2007). Whereas public institutions are responsible for helping communities cope with the physical impacts of climate change, companies must ensure that their business activities do not further exacerbate them and heighten society's exposure to danger. In practice, it can be hard to distinguish between the intersecting impacts of climate change and business activities. CSOs thus have little incentive to frame their demands in terms of climate change, as it would not only make it more technically challenging to compile evidence of mining impacts, but it would also dilute corporate responsibility for their presence.

Whereas company representatives in both the Antamina and Cerro Corona cases sought to portray climate risks as shared, organizations of civil society rejected this framing. Local communities in the Cerro Corona interpreted Gold Field's educational programs about climate adaptation as an attempt to avoid responsibility by blaming the global phenomenon of climate change without recognizing their responsibility for the double exposure of local communities to such intersecting impacts (Interview 54). Advocacy has instead focused on compelling companies to assume responsibility for their impacts on water resources.

That said, some civil society representatives have been critical of the insufficient attention given to climate adaptation. Gold Fields, for example, has submitted an EIA for extending its operations, but rural organizations have argued against approving this and other new mining projects affecting water resources in the face of climate change (Interviews 53, 57, 63). Similar concerns have been raised in the Antamina case (Interview 67). Had CSOs challenged the ways in which mining companies handle climate impacts, companies may have felt increased pressure to handle climate risks in a more transparent and inclusionary way. However, as such demands were rare, the companies in our three cases have largely abstained from engaging in public debates about climate adaptation.

Overall, civil society advocacy has concentrated on convincing companies to disclose information and compensate local communities for the negative impacts on water resources through CSR investments in water infrastructure. These demands have rarely been framed in terms of climate change. While this might be due to lack of awareness of the intersecting impacts of mining activities and climate change, the climate adaptation framing also makes it harder to ascribe responsibility to companies, a factor that provides at least a partial explanation for CSOs' avoidance of this frame. In the three cases examined, advocacy has worked to induce companies both to make CSR investments in water infrastructure and to involve local communities in participatory monitoring processes around water resources. Thus, civil society advocacy has played a crucial role in pressuring companies to integrate RBC principles. However, this pressure has been selective, as important components such as climate risk assessments and transparency were largely absent from their demands.

4.4 Conclusion

This section undertook an in-depth analysis of the extent to which, and the reasons why, companies integrate the principles of RBC within the context of three mining projects in Peru. Consistent with the descriptive findings in Section 3, all three companies primarily made CSR investments in public adaptation goods. We also showed how the regulatory and political factors privileged in our theoretical framework shaped these outcomes, although their relative importance differed. The analysis suggests that *public governance* based on soft steering mechanisms has incentivized companies to make CSR investments, but there were no regulatory controls restricting business activities that could produce socio-environmental harm in a context of climate change. Likewise, *CSO advocacy* mainly focused on demands of CSR investments but rarely pressured companies to assess and redress climate impacts in an inclusionary and transparent manner. In turn, *voluntary adaptation standards* have contributed to spread norms and learning about the four principles, but the lack of strong enforcement mechanisms have limited the effectiveness of such standards in terms of changing company behavior. In all, companies have primarily layered RBC in adaptation into existing CSR practices, rather than proactively seeking to address the risks associated with their business activities in a context of climate change through the integration of procedural principles.

5 Conclusion: The Future of Responsible Corporate Adaptation

This Element started with the premise that corporate adaptation is not merely a matter of business resilience, as climate change can exacerbate existing socio-environmental risks associated with business activities or even generate new risks. To avoid or mitigate negative impacts, corporate adaptation responses to the intersecting impacts of climate change need to be grounded in principles of RBC. Corporate adaptation to biophysical stressors is certainly not a new phenomenon, but uncertain and future climate risks complicate the socio-environmental risk landscape and raise critical questions about how general RBC principles should be operationalized in climate adaptation.

In this concluding section, we summarize our findings and discuss their broader implications. Our findings generally showed that RBC is weakly integrated into the adaptation responses of the world's largest mining companies. Companies have primarily addressed the most visible and urgent climate risks through conventional CSR approaches, with only a few sustainability frontrunners beginning to integrate procedural principles; however, none demonstrated transparency regarding the nature and scope of their adaptation responses. Based on our case studies of three mining operations in Peru, we showed that these patterns can be explained by the combination of weak public governance, voluntary adaptation standards without strong enforcement mechanisms, and the nature of civil society advocacy.

These findings from the mining sector have significant implications for ongoing debates about corporate adaptation and the relationship between private and public governance in the area of adaptation. We argue that these principles are relevant for all companies, irrespective of their size, ownership, and sector (cf. OECD 2023). That said, our findings are likely to be particularly relevant for companies whose operations depend on the intensive use of land or water, as their adaptation responses are likely to have significant consequences for the climate resilience of local ecosystems and communities. These sectors include energy, food, gas, oil, hydropower, and large-scale agriculture. Beyond these sectors, our findings are also likely to be relevant for workers' rights in all industries that are affected by climate impacts. Recent studies on the apparel sector has, for instance, emphasized the need for climate adaptation to protect workers in countries like Bangladesh, Cambodia, and Pakistan from extreme heat (Judd et al., 2023). Moreover, our findings are also generalizable to climate-vulnerable countries that depend heavily on the production of key commodities – such as soy in Brazil or lithium in Chile – where economic reliance on these industries may make governments reluctant to impose strict regulatory controls on businesses. Recognizing that the nature of adaptation challenges is highly context-dependent, our research offers critical insights into how RBC can be integrated in adaptation across different sectors and regions.

In the remainder of this section, we first summarize our main descriptive findings and discuss them in light of ongoing debates about corporate adaptation and public–private sector interactions (Section 5.1). In the second part, we highlight our main explanatory findings and further elaborate on our contributions (Section 5.2). Finally, we discuss the broader implications for the future of adaptation governance (Section 5.3).

5.1 Patterns of Responsible Business Conduct in Adaptation

Drawing on academic debates on private governance, Business and Human Rights, and climate adaptation governance, this Element has conceptualized RBC in adaptation in terms of the following principles: assessments of climate risks to society, participation, transparency, and CSR investments. We argued that these principles could play an important role in preventing and mitigating harm and enable the development of regulatory policies to address the intersecting impacts of climate change and business activities. Using both quantitative data derived from an analysis of documents produced by the thirty-seven largest mining corporations and qualitative interviews conducted with different actors involved in or affected by mining activities, we analyzed the variation in this main outcome of interest, namely, the extent to which companies integrated these principles into their adaptation responses.

The key finding from Section 3 is that only a minority of the thirty-seven observed companies integrated the RBC principles in their adaptation responses. This minority primarily did this through CSR investments in public adaptation goods rather than by addressing the procedurally oriented principles that could potentially help to prevent or mitigate the risks associated with business activities in a context of climate change. *CSR investments* can, to some extent, help to expand access to adaptation public goods, which are urgently needed in countries where governments are unable to finance the implementation of adaptation plans (Pauw and Pegels 2013; UNEP 2021). However, our qualitative findings from Peru also revealed that companies often used CSR investments to compensate local communities for the most visible climate impacts, with the underlying objective of defusing protests and legitimizing mining projects.

Regarding the *assessment of climate risks to society*, our analysis in Section 3 suggests that while a small number of the thirty-seven companies had recently begun to analyze the climate vulnerability of host communities, these assessments were underdeveloped. Companies provide little information about their methods of assessing risk or the resultant findings. Furthermore, although a few companies enabled stakeholder *participation* in processes concerning climate-related impacts, particularly related to water monitoring and governance, none had systematic procedures in place for involving local communities in the planning and implementation of adaptation activities. Likewise, there was no evidence that companies were *transparent* about their exposure to climate risks and adaptation measures. Instead, they often diverted attention from their own responsibility by portraying climate risks as isolated from their business activities.

Finally, the qualitative analysis in Section 4 revealed that the three companies we analyzed were reluctant to share information about climate risks, unless they were pressured by communities to disclose such information. This lack of transparency seriously undermines the ability of public institutions and CSOs to evaluate whether corporate adaptation responses appropriately address the full range of climate-related risks and ultimately to set up regulatory policies and hold companies accountable for their climate adaptation (in)actions.

Taken together, our findings indicate that RBC in adaptation is often selective, reactive, shortsighted, and layered into existing CSR initiatives. Instead of addressing climate risks in a transparent and inclusionary manner, companies tend to selectively respond to the most visible impacts that could threaten their social license to operate by compensating local communities through CSR initiatives and partially engaging with them through stakeholder participation processes. In contrast, companies generally took few proactive measures to identify and reduce community exposure to climate-related harms linked to their operations.

These findings make two main contributions. First, our work speaks to and builds upon a burgeoning literature on corporate adaptation from an organizational and management perspective (for an overview, see Averchenkova et al. 2016). This scholarship has deepened our understanding of firms' awareness and perceptions of climate risks (Gasbarro and Pinkse 2016; Pinkse and Gasbarro 2019; Shi and Moser 2021) and patterns of internal organizational change (Berkhout 2012). While Goldstein and colleagues (2019) observe that corporate adaptation strategies are dominated by existing risk management approaches and are thus largely devoid of ethical discourse, this literature has not yet analyzed the socioenvironmental consequences of corporate adaptation. Our study addresses this gap by mapping and explaining RBC within this increasingly critical policy and governance domain.

Second, while corporate responsibility to address any negative externality arising from core business activities such as water pollution, deforestation, land grabbing, and child labor is rather clear, the responsibility to address intersecting adaptation risks remains ambiguous. Companies often portrayed climate impacts as a "shared risk" stemming from biophysical stressors, while failing or avoiding to acknowledge how their mining activities have intensified these risks for affected communities, thereby leaving them in a situation of "double exposure" (Leichenko and O'Brien 2008). Our qualitative analysis highlights the difficulty of holding companies accountable for intersecting risks arising from both climate change and business activities, as these overlapping risks blur the boundaries of responsibility between states and corporate actors. However, existing research on corporate adaptation has largely overlooked how companies handle such intersecting risks. We have advanced the corporate adaptation literature by conceptualizing RBC in the context of such intersecting climate-related business impacts. Although this study provides an initial analysis within the mining sector, further studies are needed to examine how companies integrate RBC principles into adaptation strategies across different industries and national contexts. Moreover, while our analysis focuses on those RBC principles most relevant to the prevention of harm and the regulation of climate-related business impacts, future research could also study other key aspects of RBC in adaption, most notably corporate governance and access to remedy, to provide a more comprehensive account.

Finally, our study highlights an important insight regarding companies' perceptions of climate risks – a critical factor in shaping adaptation responses (Gasbarro and Pinkse 2016; Linnenluecke et al. 2011). Our quantitative mapping in Section 3 reveals that less than half of the thirty-seven mining companies studied recognized the societal risks associated with their business activities in the context of climate change. This finding is important, as it implies that the majority of companies are likely to continue to approach adaptation through a "business-as-usual" lens, increasing the risk of maladaptive outcomes.

5.2 Regulatory and Political Factors Shaping Responsible Business Conduct in Adaptation

Corporate adaptation is a complex sociopolitical process in which several actors govern how companies adapt to climate change. In Section 4, we turned to the second part of our theoretical framework, and in line with our expectations, our case studies of mining operations in Peru demonstrated that the weak integration of the RBC principles in adaptation can be attributed to a combination of three main factors: public governance, private voluntary standards, and civil society advocacy.

Our analysis showed that *public adaptation governance* in Peru is still in its early stages and imposes no legal requirements on companies to comply with the four principles. Instead of monitoring, restricting, and prohibiting business activities in climate-impacted areas, public authorities – both at the national and subnational levels – have focused on mobilizing companies through multistakeholder dialogues and public–private partnerships to incentivize them to make CSR investments in public adaptation goods. The need for private funding in the adaptation space creates opportunities for companies to showcase their problem-solving capacity in helping society in coping with climate change. In contrast to the procedurally oriented principles, CSR investments present a "win-win situation" (cf. Auld et al. 2008: 415) for companies interested in securing their social license to operate.

Regarding the procedural principles, public institutions have developed voluntary guidelines for how to integrate climate risks into EIAs, which stipulate that companies should be transparent about climate risks and ensure consultation with local communities about their adaptation responses, but in the absence of binding rules, the uptake of such guidelines has been limited. In general, we found that the Peruvian state did impose few requirements or incentives for companies to assess societal climate risks, be transparent, and ensure stakeholder participation in their adaptation responses.

While public governance efforts have focused on promoting CSR investments, *voluntary adaptation standards* have disseminated norms and knowledge related to all four principles. Our analysis in Section 3 indicates that ICMM members have adopted a more explicit discourse regarding mining companies' responsibility to support – or at least not undermine – societal climate resilience, in contrast to nonmembers. At the same time, Section 4 showed that while ICMM standards have contributed to capacity building, adaptation is only partially included into audited performance indicators. Given weak enforcement mechanisms, the way such standards have shaped company practices at the operational level is uneven and highly dependent on staff commitment.

Consistent with earlier studies documenting the limited uptake of voluntary standards, particularly when private standards are not countervailed with a legislative threat or CSO advocacy (Bartley 2018b; Börzel and Risse 2010), we find that voluntary adaptation standards have been insufficient to make companies effectively comply with RBC in adaptation. This might be a result of the ambiguous meaning of "adaptation," which makes it challenging to develop measurable indicators and monitor them effectively. For these reasons, instituting stringent adaptation standards is likely to be challenging (cf. Thompson et al. 2021).

Lastly, *civil society advocacy* has, similarly to public governance, focused on demands of CSR investments in public adaptation goods. We found only a few examples of CSOs pressuring companies to integrate the procedurally oriented principles. Rather, the results suggest that CSR investments in public adaptation goods have diminished the incentives for local communities to push for public debates on the adverse impacts of mining in adaptation. CSR investments can thus crowd out the adoption of procedural principles that would be critical for preventing socio-environmental harms, mitigating their impacts, and coordinate public and private governance of intersecting climate-related business impacts on society. These findings are in line with previous studies that have highlighted the adverse effects of CSR in the context of mining (e.g., Amengual 2018; Bebbington 2010; Gustafsson 2018).

These findings support our expectations and thus make an important contribution to the growing literature on corporate adaptation. This literature has highlighted how different types of internal and external factors, such as organizational structures and climate risk exposure, have shaped corporate adaptation responses (Averchenkova et al. 2016; Berkhout 2012; Pinkse and Gasbarro 2019). However, to date they have not studied to what extent such factors are also relevant in relation to RBC. Moreover, existing literature has not analyzed how the three regulatory and political factors that we focused on shape firms' responses to climate risks. Against this backdrop, we identified significant barriers associated with each of these factors, making the prospects for changing company behavior appear bleak, at least in the short term. Further research is needed to explore how these regulatory and political factors can be strengthened to encourage companies to comply more meaningfully with RBC principles. While we have focused on public governance, civil society advocacy, and voluntary adaptation standards, future research could study how internal and external factors combine in shaping RBC in adaptation.

Our findings also advance ongoing debates about the relationship between private and public climate governance. This literature has recognized public regulation as an important factor shaping corporate decarbonization strategies (Newell and Daley 2024; Sakhel 2017). By contrast, the adaptation literature has concentrated primarily on public adaptation governance and, apart from a few notable exceptions (Klein et al. 2018; Pauw and Pegels 2013), largely overlooked the role of private companies. This Element contributes by presenting a deeper understanding of the opportunities and challenges that often accompany the use of soft steering to incentivize responsible business behavior. Soft steering has enabled companies to strategically position themselves in the adaptation space. For example, previous research has shown that soft steering has allowed large agricultural producers in Brazil to shape adaptation policy and resulted in a market-liberal adaptation agenda that enables the intensification of production in a context of climate change (Gustafsson et al. 2024; Milhorance et al. 2022a).

Generally, soft steering approaches can enhance corporate motivation and capacity to integrate the RBC principles. However, if such approaches are entirely based on voluntary contributions, without being backed up by any legal requirements or strong domestic institutions, their effect is likely to be limited (Börzel et al. 2012; Eberlein 2019). Our study shows that the construction of adaptation policies capable of balancing competing demands depends on strict legal requirements that climate risks are considered in sectoral tools and on robust public monitoring systems to make it possible for state agencies to assess climate risks and the impacts of business activities. Absent these conditions, it is unlikely that public institutions will have the capacity to countervail corporate influence in adaptation governance. There is a need for more research on the conditions under which governments are willing and able to adopt and enforce corporate regulatory controls in the area of adaptation.

5.3 The Future of Adaptation Governance

Strengthening public adaptation governance is critical to addressing the escalating and urgent impacts of climate change worldwide, raising important questions about the extent and nature of corporate involvement in this process. In this concluding section, we reflect on the dual role of companies in shaping public adaptation governance, highlighting how they can reinforce or undermine institutional responses to climate change.

On the one hand, there has been a hope among global policymakers that private sector investments can be used as a vehicle for generating resources to build institutional capacity and implement adaptation plans (UNEP 2021). Limited resources and bureaucratic capacity stand as significant barriers to reducing the implementation gap in adaptation, particularly in regions of the Global South (Milhorance et al. 2022b). There, companies could contribute to the expansion of state capacity in adaptation. In addition, because climate risks are a cross-cutting policy issue implicating the core of energy, social, and industrial policy, among others, the inclusion of the private sector is needed to foster more integrated solutions to climate impacts.

On the other hand, it is important to acknowledge the significant risk that corporations may undermine ambitious adaptation governance. Adaptation is a political process wherein firms bring their own policy preferences and priorities and may use their structural power to bend the adaptation agenda in their favor. Governments and public authorities aiming to foster more sustainable corporate adaptation practices must carefully construct adaptation policies to balance potentially competing demands, particularly when business activities involve the intensive use of natural resources, as in mining and large-scale agriculture. Developing such policies requires not only strong pro-regulatory political coalitions but also robust public monitoring systems to be able to evaluate the intersecting impacts of business activities and climate change. However, the lack of transparency regarding corporate climate risk exposure and adaptation responses is likely to make it difficult for state institutions to assess the adequacy of corporate adaptation strategies.

CSR investments have enabled companies to position themselves as important agents in the quest to resolve the pressing problems of adaptation while rendering surrounding societies dependent on corporate benevolence. Existing research, particularly from the mining sector, has shown that CSR programs have often reduced pressure for institutional reform, created divisions among CSOs, and shut down public debate on the negative impacts of mining activities (Amengual 2018; Bebbington 2010; Gustafsson 2018). This suggests that CSR investments might serve as an important tool for companies to undermine ambitious adaptation governance in areas where corporations view such policies as a threat to their business activities. For example, they are likely to oppose policies that restrict or prohibit business activities in climate-affected areas. Our findings highlight the need for scholars to critically assess the risks associated with delegating authority to companies in the adaptation space.

Overall, in regions where vulnerable groups are exposed to the intersecting impacts of climate change and business activities, companies have emerged as important agents in adaptation governance. While companies have often compensated local communities for the visible impacts of climate change through business-as-usual practices, our findings underscore the urgent need to strengthen the capacity of both public institutions and CSOs to hold companies accountable for how they handle the profound implications of such risks to society and ecosystems.

References

Abbott, K. and Snidal, D. (2009). Strengthening international regulation through transnational new governance: Overcoming the orchestration deficit. *Vanderbilt Journal of Transnational Law*, **42**, 501–578. https://ssrn.com/abstract=1403517.

Adger, W. N., Brown, I., and Surminski, S. (2018). Advances in risk assessment for climate change adaptation policy. *Philosophical Transactions of the Royal Society A: Mathematical, Physical and Engineering Sciences*, **376**(2121), 20180106. https://doi.org/10.1098/rsta.2018.0106.

Alliance for Water Stewardship (2022). AWS *Strategy 2022–2030: Accelerating Impact*. https://a4ws.org/download/aws-strategy-2022-2030/.

Amengual, M. (2018). Buying stability: The distributive outcomes of private politics in the Bolivian mining industry. *World Development*, **104**, 31–45. https://doi.org/10.1016/j.worlddev.2017.11.008.

Anglo American (n.d). Mesa de Dialogo. https://peru.angloamerican.com/es-es/quellaveco/mesa-de-dialogo.

Anglo American (2018). Sustainability Report 2018: Unlocking our Full Potential, Disciplined Growth for a Sustainable Future. https://southafrica.angloamerican.com/~/media/Files/A/Anglo-American-Group-v5/South-Africa/media/documents/anglo-sustainability-report-2018.pdf.

Anglo American (2019). *Anglo American CDP Water Security Questionnaire 2019*. www.angloamerican.com/~/media/Files/A/Anglo-American-Group/PLC/sustainability/approach-and-policies/sustainability/performance/anglo-american-water-security-response-2019.pdf.

Anglo American (2022). *Climate Change Report 2022*. www.angloamerican.com/~/media/Files/A/Anglo-American-Group-v5/PLC/investors/annual-reporting/2022/climate-change-report-2022.pdf.

AngloGold Ashanti (2020). *Sustainability Report 2019*. www.aga-reports.com/19/sr.

Antamina (2017). Ministerio de Agricultura y Riego y Antamina concretan primera obra por impuestos en el sector agrario público. March 9. www.antamina.com/noticias/ministerio-agricultura-riego-antamina-concretan-primera-obras_por_impuestos-sector-agrario-publico/.

Antamina (2020). *Sustainability Report 2020*. www.antamina.com/wp-content/uploads/2021/09/sustainability-report-antamina2020.pdf.

Antofagasta (2019). *Developing Mining for a Better Future: Annual Report and Financial Statements 2019*. www.antofagasta.co.uk/media/3772/antofagasta-2019-annual-report.pdf.

Arellano-Yanguas, J. (2018). Extractive industries and regional development: Lessons from Peru on the limitations of revenue devolution to producing regions. *Regional & Federal Studies*, 1–25. https://doi.org/10.1080/13597566.2018.1493461.

Auld, G., Bernstein, S., and Cashore, B. (2008). The new corporate social responsibility. *Annual Review of Environment and Resources*, **33**, 413–435. https://doi.org/10.1146/annurev.energy.32.053006.141106.

Averchenkova, A., Crick, F., Kocornik-Mina, A., Leck, H., et al. (2016). Multinational and large national corporations and climate adaptation: Are we asking the right questions? *WIREs Climate Change*, **7**(4), 517–536. https://doi.org/10.1002/wcc.402.

Balsiger, P. (2014). *The Fight for Ethical Fashion – The Origins and the Interactions of Clean Clothes Campaign*. Aldershot: Ashgate.

Barrick (2022). Barrick Sustainability Report 2022. www.barrick.com/English/sustainability/default.aspx.

Bartley, T. (2018a). Transnational corporations and global governance. *Annual Review of Sociology*, **44**, 145–165. https://doi.org/10.1146/annurev-soc-060116-053540.

Bartley, T. (2018b). *Rules without Rights: Land, Labor, and Private Authority in the Global Economy*. Oxford: Oxford University Press.

Bassett, T. and Fogelman, C. (2013). Déjà vu or something new? The adaptation concept in the climate change literature. *Geoforum*, **48**, 42–53. https://doi.org/10.1016/j.geoforum.2013.04.010.

Bebbington, A. (2010). Extractive industries and stunted states: Conflict, responsibility and institutional change in the Andes. In R. K. Ravi and R. D. Lipschutz, eds., *Corporate Social Responsibility, Discourses, Practices and Perspectives*. London: Palgrave Macmillan, pp. 97–115.

Bebbington, A. and Bury, J. (2009). Institutional challenges for mining and sustainability in Peru. *Proceedings of the National Academy of Sciences*, **106**(41), 17296–17301. https://doi.org/10.1073/pnas.0906057106.

Begum, R. A., Lempert, R., Ali, E., Benjaminsen, T. A., et al. (2022). Point of departure and key concepts. In H.-O. Pörtner, D. C. Roberts, M. Tignor, et al., eds., *Climate Change 2022: Impacts, Adaptation and Vulnerability. Contribution of Working Group II to the Sixth Assessment Report of the Intergovernmental Panel on Climate Change*. Cambridge, UK and New York, USA: Cambridge University Press, pp. 121–196.

Berkhout, F. (2012). Adaptation to climate change by organizations. *WIREs Climate Change*, **3**(1), 91106. https://doi.org/10.1002/wcc.154.

Bernaz, N. (2016). *Business and Human Rights: History, Law and Policy – Bridging the Accountability Gap*. London: Routledge.

Börzel, T. A. and Risse, T. (2010). Governance without a state: Can it work? *Regulation & Governance*, **4**(2), 113–134. https://doi.org/10.1111/j.1748-5991.2010.01076.x.

Börzel, T. A., Hönke, J., and Thauer, C. R. (2012). Does it really take the state? *Business and Politics*, **14**(3), 1–34. https://doi.org/10.1515/bap-2012-0023.

Boulianne, S., Lalancette, M., and Ilkiw, D. (2018). School strike 4 climate: Social media and the international youth protest on climate change. *Media and Communication*, **8**(2), 208–218. https://doi.org/10.17645/mac.v8i2.2768.

Bury, J., Mark, B., Carey, M., McKenzie, J., et al. (2013). New geographies of water and climate change in Peru. *Annals of the Association of American Geographers*, **103**(2), 363–374. https://doi.org/10.1080/00045608.2013.754665.

Carpenter, R. C. (2007). Setting the advocacy agenda: Theorizing issue emergence and nonemergence in transnational advocacy networks. *International Studies Quarterly*, **51**(1), 99–120. https://doi.org/10.1111/j.1468-2478.2007.00441.x.

Cashore, B., Knudsen, J. S., Moon, J., and van der Ven, H. (2021). Private authority and public policy interactions in global context: Governance spheres for problem solving. *Regulation & Governance*, **15**(4), 1166–1182. https://doi.org/10.1111/rego.12395.

Castro Pereira, J. (2023). Peru at the UNFCCC: Explaining the country's foreign climate policy. *Climate Policy*, **23**(2), 212–225. https://doi.org/10.1080/14693062.2022.2100733.

Cenepred (2023). Sistema de Informacion para la Gestion de Riesgo de Desastres (SIGRID). Geospatial Map. https://sigrid.cenepred.gob.pe/sigridv3/.

CooperAcción (2021). Antamina y los Conflictos de Baja Intensidad. CooperAcción, July 14. https://cooperaccion.org.pe/opinion/antamina-y-los-conflictos-de-baja-intensidad/.

CooperAcción (2022). Conflicto sobre Licencia de Agua a Quellaveco. CooperAcción, October 27. https://cooperaccion.org.pe/conflicto-sobre-licencia-de-agua-a-quellaveco-boletin-amp-oct-2022/.

de Bakker, F. G., Rasche, A., and Ponte, S. (2019). Multi-stakeholder initiatives on sustainability: A cross-disciplinary review and research agenda for business ethics. *Business Ethics Quarterly*, **29**(3), 343–383. https://doi.org/10.1017/beq.2019.10.

de Moor, J. (2022). Prioritizing adaptation and mitigation in the climate movement: Evidence from a cross-national protest survey of the Global Climate Strike, 2019. *Mitigation and Adaptation Strategies for Global Change*, **27**(4), 1–19. https://doi.org/10.1007/s11027-022-10003-y.

Dellmuth, L. and Bloodgood, E. A. (2023). Advocacy group effects in global governance: A theoretical framework. In L. Dellmuth and E. A. Bloodgood, eds., *Advocacy Group Effects in Global Governance*. Cham: Palgrave Macmillan. https://doi.org/10.1007/978-3-031-27864-8_1.

Dellmuth, L. and Gustafsson, M. T. (2023). Legitimacy in the trans-scalar governance of climate adaptation. *npj Climate Action*, **2**, 2. https://doi.org/10.1038/s44168-023-00036-7.

Depoorter, C. and Marx, A. (2024). Fostering compliance with voluntary sustainability standards through institutional design: An analytic framework and empirical application. *Regulation & Governance*, **18**(4), 1132–1152. https://doi.org/10.1111/rego.12573.

Dolšak, N. and Prakash, A. (2018). The politics of climate change adaptation. *Annual Review of Environment and Resources*, **43**, 317–341. https://doi.org/10.1146/annurev-environ-102017-025739.

Dupuis, J. and Biesbroek, R. (2013). Comparing apples and oranges: The dependent variable problem in comparing and evaluating climate change adaptation policies. *Global Environmental Change*, **23**(6), 1476–1487. https://doi.org/10.1016/j.gloenvcha.2013.07.022.

Eberlein, B. (2019). Who fills the global governance gap? Rethinking the roles of business and government in global governance. *Organization Studies*, **40**(8), 1125–1145. https://doi.org/10.1177/0170840619847720.

Equator Principles (2020). *The Equator Principles: July 2020*. https://equator-principles.com/app/uploads/The-Equator-Principles_EP4_July2020.pdf.

Eriksen, S., Schipper, E., Scoville-Simonds, M., Vincent, K., et al. (2021). Adaptation interventions and their effect on vulnerability in developing countries: Help, hindrance or irrelevance? *World Development*, **141**, 105383. https://doi.org/10.1016/j.worlddev.2020.105383.

Feindouno, S., Guillaumont, P., and Simonet, C. (2020). The physical vulnerability to climate change index: An index to be used for international policy. *Ecological Economics*, **176**, 106752. https://doi.org/10.1016/j.ecolecon.2020.106752.

FERDI (2022). *The Physical Vulnerability to Climate Change Index (PVCCI)*. https://ferdi-indicators.shinyapps.io/PVCCI/.

Franks, D., Davis, R., Bebbington, A., and Scurrah, M. (2014). Conflict translates environmental and social risk into business costs. *Proceedings of the National Academy of Sciences*, **111**(21), 7576–7581. https://doi.org/10.1073/pnas.1405135111.

Freeport-McMoRan (2021). *Climate Report 2021*. www.fcx.com/sites/fcx/files/documents/sustainability/2021-Climate-Report.pdf.

Gardner, T., Benzie, M., Börner, J., Dawkins, E., et al. (2019). Transparency and sustainability in global commodity supply chains. *World Development*, **121**, 163–177. https://doi.org/10.1016/j.worlddev.2018.05.025.

Gasbarro, F. and Pinkse, J. (2016). Corporate adaptation behaviour to deal with climate change: The influence of firm-specific interpretations of physical climate impacts. *Corporate Social Responsibility and Environmental Management*, **23**(3), 179–192. https://doi.org/10.1002/csr.1374.

Glencore (2021). *Pathway to Net Zero 2021 Progress Report*. www.glencore.com/.rest/api/v1/documents/12b9c4417f45c969007f6e09ebf2ca67/2021-Climate-Change-Report-+%282%29.pdf.

Gobierno Regional de Ancash (2017). *Estrategia Regional de Cambio Climático de Ancash: Resumen segunda edición*. https://proyectoglaciares.care.org.pe/wp-content/uploads/2017/11/ERCC-en-baja-FINAL-APROBADO.pdf.

Gobierno Regional de Cajamarca (2016). *Estrategia Regional Frente al Cambio Climático – Cajamarca al 2030*. https://siar.regioncajamarca.gob.pe/documentos/estrategia-regional-frente-cambio-climatico-cajamarca-2030.

Goldstein, A., Turner, W., Gladstone, J., and Hole, D. (2019). The private sector's climate change risk and adaptation blind spots. *Nature Climate Change*, **9**(1), 18–25. https://doi.org/10.1038/s41558-018-0340-5.

Gold Fields (2019). *Integrated Annual Report 2019*. www.goldfields.com/pdf/investors/integrated-annual-reports/2019/full-iar.pdf.

Gold Fields (2021a). *Report to Stakeholders 2021*. www.goldfields.com/pdf/investors/integrated-annual-reports/2021/gold-fields-report-to-stakeholders-2021.pdf.

Gold Fields (2021b). *Integrated Report 2021*. www.goldfields.com/reports/annual-report-2021/pdf/iar-2021.pdf.

Gold Fields (2021c). *Climate Change Report 2021*. www.goldfields.com/pdf/investors/integrated-annual-reports/2021/gold-fields-tcfd-report-2021-updated.pdf.

Gold Fields (2022). *Climate Change Report 2022*. www.goldfields.com/pdf/investors/integrated-annual-reports/2022/ccr-2022-report.pdf.

Goujon, M., Santoni, O., and Wagner, L. (2022). The Physical Vulnerability to Climate Change Index computed at the sub-national level. FERDI Working Paper P305.

Grabs, J. (2020). Assessing the institutionalization of private sustainability governance in a changing coffee sector. *Regulation & Governance*, **14**(2), 362–387. https://doi.org/10.1111/rego.12212.

Grabs, J., et al. (2026). Resilience of what and for whom? Climate change mitigation and adaptation in the global, Ethiopian, and Tanzanian coffee sectors. *World Development*, **200**(2026). https://doi.org/10.1016/j.worlddev.2025.107299.

Gunningham, N., Kagan, R. A., and Thornton, D. (2004). Social license and environmental protection: Why businesses go beyond compliance. *Law & Social Inquiry*, **29**(2), 307–341. https://doi.org/10.1086/423681.

Gustafsson, M. T. (2018). *Private Politics and Peasant Mobilization: Mining in Peru*. Basingstoke: Palgrave Macmillan.

Gustafsson, M. T. and Schilling-Vacaflor, A. (2022). Indigenous Peoples and multiscalar environmental governance: The opening and closure of participatory spaces. *Global Environmental Politics*, **22**(2), 70–94. https://doi.org/10.1162/glep_a_00642.

Gustafsson, M-T. and Scurrah, M. (2023). Subnational governance strategies at the extractive frontier: Collaboration and conflict in Peru. *Territory, Politics, Governance*, **11**(1), 1–18. https://doi.org/10.1080/21622671.2020.1840425.

Gustafsson, M. T., Rodriguez-Morales, J. E., and Dellmuth, L. M. (2022). Private adaptation to climate risks: Evidence from the world's largest mining companies. *Climate Risk Management*, **35**, 100386. https://doi.org/10.1016/j.crm.2021.100386.

Gustafsson, M-T., Schilling-Vacaflor, A., and Pahl-Wostl, C. (2024). Governing transnational water and climate risks in global supply chains. *Earth System Governance*, **21**, 100217. https://doi.org/10.1016/j.esg.2024.100217.

Hannah, L., Roehrdanz, P., Ikegami, M., and Hijmans, R. (2013). Climate change, wine, and conservation. *Proceedings of the National Academy of Sciences*, **110**(17), 6907–6912. https://doi.org/10.1073/pnas.1210127110.

Higham, I. (2019). UN guiding principles on business and human rights. In O. E. Olsen, K. V. Juhl, P. H. Lindoe, and O. A. Engen, eds., *Standardization and Risk Governance*. London: Routledge, pp. 217–234.

Hopkins, A. and Kemp, D. (2021). *Credibility Crisis: Brumadinho and the Politics of Mining Industry Reform*. Amsterdam: Wolters Kluwer.

ICMM (2016). *Mitigating GHG Emissions and Building Resilience*. International Council on Mining and Metals. www.icmm.com/en-gb/our-work/environmental-resilience/climate-change/mitigate-ghg-emissions.

ICMM (2017). *Position Statement: Water Stewardship*. www.icmm.com/website/publications/pdfs/mining-principles/position-statements_water-stewardship.pdf?cb=59905.

ICMM (2019). *Adaptation to a Changing Climate: Building Resilience in the Mining and Metals Industry*. www.icmm.com/en-gb/guidance/environmental-stewardship/2019/adapting-to-a-changing-climate.

ICMM (2020). *Global Industry Standard on Tailings Management*. www.icmm.com/en-gb/our-principles/tailings/global-industry-standard-on-tailings-management.

ICMM (2021). *Climate Change: Position Statement*. www.icmm.com/website/publications/pdfs/mining-principles/position-statements_climate-change.pdf?cb=59897.

ICMM (2022). *Role of Mining in National Economies: Mining Contribution Index* (6th ed.). London, UK: ICMM. Retrieved from www.icmm.com/en-gb/research/social-performance/2022/role-of-mining-in-national-economies; https://pimcore.icmm.com/website/publications/pdfs/social-performance/2022/research_mci-6-ed.pdf?cb=16134.

IPCC (2022). *Climate Change 2022: Impacts, Adaptation, and Vulnerability. Contribution of Working Group II to the Sixth Assessment Report of the Intergovernmental Panel on Climate Change*. H.-O. Pörtner, D. C. Roberts, and M. Tignor, et al., eds. Cambridge: Cambridge University Press.

Jaskoski, M. (2014). Environmental licensing and conflict in Peru's mining sector: A path-dependent analysis. *World Development*, **64**, 873–883.

Jernnäs, M. and Lövbrand, E. (2022). Accelerating climate action: The politics of nonstate actor engagement in the Paris regime. *Global Environmental Politics*, **22**(3), 38–58. https://doi.org/10.1162/glep_a_00660.

Judd, J., Bauer, A., Kuruvilla, S., and Stephanie, W. (2023). *Higher Ground? Report 1: Fashion's Climate Breakdown and Its Effect for Workers*. Retrieved from: www.ilr.cornell.edu/global-labor-institute/higher-ground-fashions-climate-breakdown.

Klein, J., Araos, M., Karimo A., Heikkenen, M., et al. (2018). The role of the private sector and citizens in urban climate change adaptation: Evidence from a global assessment of large cities. *Global Environmental Change*, **53**, 127–136. https://doi.org/10.1016/j.gloenvcha.2018.09.012.

Klein, J., Käyhkö, J., Räsänen, A., Groundstroem, F., et al. (2022). Climate risk perception, management, and adaptation in the Nordic mining sector. *The Extractive Industries and Society*, **10**, 101092. https://doi.org/10.1016/j.exis.2022.101092.

Kramarz, T., Mason, M., and Partzsch, L. (2023). Proxy-led accountability for natural resource extraction in rentier states. *Environmental Politics*, **32**(1), 113–134. https://doi.org/10.1080/09644016.2022.2044219.

Lambin, E. and Thorlakson, T. (2018). Sustainability standards: Interactions between private actors, civil society, and governments. *Annual Review of Environment and Resources*, **43**, 369–393. https://doi.org/10.1146/annurev-environ-102017-025931.

Lawrence, R. and Kløcker Larsen, R. K., (2017). The politics of planning: Assessing the impacts of mining on Sami lands. *Third World Quarterly*, **38**(5), 1164–1180. https://doi.org/10.1080/01436597.2016.1257909.

LeBaron, G., Lister, J., and Dauvergne, P. (2017). Governing global supply chain sustainability through the ethical audit regime. *Globalizations*, **14**(6), 958–975. https://doi.org/10.1080/14747731.2017.1304008.

Lèbre, É., Stringer, M., Svobodova, K., Owen, J., et al. (2020). The social and environmental complexities of extracting energy transition metals. *Nature Communications*, **11**(1), 4823. https://doi.org/10.1038/s41467-020-18661-9.

Leichenko, R. M. and O'Brien, K. L. (2008). *Environmental Change and Globalization: Double Exposures*. Oxford: Oxford University Press.

Leme da Silva, A., Eloy, L., Oliveira, K., Coelho Filho, O., et al. (2023). Environmental policy reform and water grabbing in an agricultural frontier in the Brazilian Cerrado. *IDS Bulletin*, **54**(1), 89–101. https://doi.org/10.19088/1968-2023.107.

Linnenluecke, M., Stathakis, A., and Griffiths, A. (2011). Firm relocation as adaptive response to climate change and weather extremes. *Global Environmental Change*, **21**(1), 123–133. https://doi.org/10.1016/j.gloenvcha.2010.09.010.

Lutheran World Relief (2017). Where we Work: Peru. https://lwr.org/where-we-work/peru.

Macchi, C. (2022). *Business, Human Rights and the Environment: The Evolving Agenda*. The Hague: T.M.C. Asser Press.

Macklin, M. G., Thomas, C., Mudbhatkal, H., and Brewer, P., et al. (2023). Impacts of metal mining on river systems: A global assessment. *Science*, **381**(6664), 1345–1350. https//10.1126/science.adg6704.

Marx, A., Depoorter, C., Fernandez de Cordoba, R., Verma, R., et al. (2024). Global governance through voluntary sustainability standards: Developments, trends and challenges. *Global Policy*, **15**(4), 708–728. https://doi.org/10.1111/1758-5899.13401.

Mayembe, R., Simpson, N. P., Rumble, O., and Norton, M. (2023). Integrating climate change in environmental impact assessment: A review of requirements across 19 EIA regimes. *The Science of the Total Environment*, **869**, 161850. https://doi.org/10.1016/j.scitotenv.2023.161850.

Mbanyele, W. and Muchenje, L. (2022). Climate change exposure, risk management and corporate social responsibility: Cross-country evidence. *Journal of Multinational Financial Management*, **66**, 100771. https://doi.org/10.1016/j.mulfin.2022.100771.

Milhorance, C., Sabourin, E., Chechi, L., and Mendes, P. (2022a). The politics of climate change adaptation in Brazil: Framings and policy outcomes for the rural sector. *Environmental Politics*, **31**(2), 183–204. https://doi.org/10.1080/09644016.2021.1907097.

Milhorance, C., Howland, F., Sabourin, E., and Le Coq, J. (2022b). Tackling the implementation gap of climate adaptation strategies: Understanding policy translation in Brazil and Colombia. *Climate Policy*, 22(9–10), 1113–1129. https://doi.org/10.1080/14693062.2022.2085650.

Ministerio del Ambiente (MINAM) (n.d.). *Hoja de Ruta Actualizada para el Involucramiento del Sector Privado y Financiero en Nuestro Desafío Climático o NDC al 2030*. https://cdn.www.gob.pe/uploads/document/file/2547083/Hoja%20de%20Ruta%20del%20sector%20privado%20y%20financiero.pdf.pdf.

Ministerio del Ambiente (MINAM) (2016). *Resolución Ministerial No. 272-2016*. https://cdn.www.gob.pe/uploads/document/file/155812/RM-N_-272-2016-MINAM.pdf?v=1532448237.

Ministerio del Ambiente (MINAM) (2021). *Plan Nacional de Adaptación al Cambio Climático*. www.gob.pe/institucion/minam/normas-legales/1955977-096-2021-minam.

Ministerio de Energía y Minas (MINEM) (2022). *Boletin Estadístico Minero: Edición N° 12-2021*. www.gob.pe/institucion/minem/informes-publicaciones/2729736-boletin-estadistico-minero-diciembre-2021.

Murguía, D. I. and Bastida, A. E. (2024). The elephant in the mine: Why voluntary sustainability standards are insufficient to ensure responsible mining. *The Extractive Industries and Society*, **19**, 101485. https://doi.org/10.1016/j.exis.2024.101485.

National Business Initiative (2018). The uMhlathuze Water Stewardship Partnership – changing the Water Landscape together. www.nbi.org.za/focus-areas/environmental-sustainability/water/umhlathuze-water-stewardship-partnership/.

ND-GAIN (2018). Country Rankings: Country Index, Vulnerability, and Readiness. https://gain-new.crc.nd.edu/ranking.

Newell, P. (2008). Civil society, corporate accountability and the politics of climate change. *Global Environmental Politics*, **8**(3), 122–153. https://doi.org/10.1162/glep.2008.8.3.122.

Newell, P. and Daley, F. (2024). Supply-side climate policy: A new frontier in climate governance. *WIREs Climate Change*, e909. https://doi.org/10.1002/wcc.909.

Newmont (2019). *CDP Climate Change Response 2019*.

Newmont (2022). *Climate Report 2022*. https://s24.q4cdn.com/382246808/files/doc_downloads/2023/05/Newmont-2022-Climate-Report.pdf.

Nightingale, A., Eriksen, S., Taylor, M., Forsyth, T., et al. (2020). Beyond technical fixes: Climate solutions and the great derangement. *Climate and Development*, **12**(4), 343–352. https://doi.org/10.1080/17565529.2019.1624495.

Odell, S. D. (2021). Desalination in Chile's mining regions: Global drivers and local impacts of a technological fix to hydrosocial conflict. *Journal of Cleaner Production*, **323**, 129104. https://doi.org/10.1016/j.jclepro.2021.129104.

Odell, S. D., Bebbington, A., and Frey, K. E. (2018). Mining and climate change: A review and framework for analysis. *The Extractive Industries and Society*, **5**(1), 201–214. https://doi.org/10.1016/j.exis.2017.12.004.

OECD (2023). *OECD Guidelines for Multinational Enterprises on Responsible Business Conduct*. Paris: OECD Publishing. https://doi.org/10.1787/81f92357-en.

OECD (n.d.). About Responsible Business Conduct. https://mneguidelines.oecd.org/about.htm.

OHCHR (2011). *Guiding Principles of Business and Human Rights*. New York: United Nations. www.ohchr.org/en/publications/reference-publications/guiding-principles-business-and-human-rights.

Pareja, C., Xavier, A., and Daitch, S. (2019). *Participatory Environmental Monitoring Committees in Mining Contexts: Lessons from Nine Case Studies in Four Latin American Countries*. New York: UNDP.

Pauw, P. and Pegels, A. (2013). Private sector engagement in climate change adaptation in least developed countries: An exploration. *Climate Development*, **5**(4), 257–267. https://doi.org/10.1080/17565529.2013.826130.

Phillips, J. (2016). Climate change and surface mining: A review of environment-human interactions & their spatial dynamics. *Applied Geography*, **74**, 95–108. https://doi.org/10.1016/j.apgeog.2016.07.001.

Pinkse, J. and Gasbarro, F. (2019). Managing physical impacts of climate change: An attentional perspective on corporate adaptation. *Business & Society*, **58**(2), 333–368. https://doi.org/10.1177/0007650316648688.

Potoski, M. and Prakash, A. (2013). Green clubs: Collective action and voluntary environmental programs. *Annual Review of Political Science*, **16**(1), 399–419. https://doi.org/10.1146/annurev-polisci-032211-211224.

Pulver, S. and Benney, T. (2013). Private-sector responses to climate change in the Global South. *WIREs Climate Change*, **4**(6), 479–496. https://doi.org/10.1002/wcc.240.

Purdon, M. and Thornton, P. (2019). Research methodology for adaptation policy analysis: Embracing the eclectic messy centre. In E. Keskitalo and B. Preston, eds., *Research Handbook on Climate Change Adaptation Policy*. London: Edward Elgar Publishing, pp. 157–197.

PWC (2019). *Mine 2019 Resourcing the Future*. Melbourne, Australia: PwC.

Rattis, L., Brando, P. M., Macedo, M. N., Spera, S., et al. (2021). Climatic limit for agriculture in Brazil. *Nature Climate Change*, **11**(12), 1098–1104. https://doi.org/10.1038/s41558-021-01214-3.

Rio Tinto (2016). *Rio Tinto Minera Peru: La Granja Project*. London: ICMM.

Rio Tinto (2018). *Our Approach to Climate Change*. www.riotinto.com/en/invest/reports/climate-change-report.

Rio Tinto (2021). *Our Approach to Climate Change*. www.riotinto.com/en/invest/reports/climate-change-report.

RMI (2022). Thematic Results: Environmental Responsibility, RMI Report 2022. https://2022.responsibleminingindex.org/en/results/thematic/1455.

Sakhel, A. (2017). Corporate climate risk management: Are European companies prepared? *Journal of Cleaner Production*, **165**, 103–118. https://doi.org/10.1016/j.jclepro.2017.07.056.

Shi, L. and Moser, S. (2021). Transformative climate adaptation in the United States: Trends and prospects. *Science*, **372**(6549). https://doi.org/10.1126/science.abc8054.

Shi, L., Chu, E., Anguelovski, I., Aylett, A., et al. (2016). Roadmap towards justice in urban climate adaptation research. *Nature Climate Change*, **6**(2), 131–137. https://doi.org/10.1038/nclimate2841.

South 32 (2022). Sustainable Development Report 2022. www.south32.net/docs/default-source/annual-reporting-suite/2022/sustainable-development-report-2022.pdf?sfvrsn=c351fe82_3.

Strindevall, I., Gustafsson, M. T., and Dellmuth, L. (2022). *Climate risks and community resilience in the mining sector.* Policy Report for Mistra Geopolitics. www.mistra-geopolitics.se/publications/climate-risks-and-community-resilience-in-the-mining-sector/.

TCFD (2017). *Implementing the Recommendations of the Task Force on Climate-Related Financial Disclosures.* https://assets.bbhub.io/company/sites/60/2021/10/FINAL-2017-TCFD-Report.pdf.

Teck (2022). *Climate Change Outlook 2021, Teck's TCFD-Aligned Report.* www.teck.com/media/Teck_Climate_Change_Outlook_2021.pdf.

The Moscow Times (2020). Russia admits to "World's Largest" Arctic oil spill. December 24. www.themoscowtimes.com/2020/12/24/russia-admits-to-worlds-largest-arctic-oil-spill-a72477.

Thompson, L., Davis, M., Mosley-Thompson, E., Porter, S., et al. (2021). The impacts of warming on rapidly retreating high-altitude, low-latitude glaciers and ice core-derived climate records. *Global and Planetary Change*, **203**, 103538. https://doi.org/10.1016/j.gloplacha.2021.103538.

Thorlakson, T., Hainmueller, J., and Lambin, E. F.(2018a). Improving environmental practices in agricultural supply chains: The role of company standards. *Global Environmental Change*, **48**, 32–42.

Thorlakson, T., De Zegher, J. F., and Lambin, E. F. (2018b). Companies' contribution to sustainability through global supply chains. *Proceedings of the National Academy of Sciences*, **115**(9), 2072–2077.

Tompkins, E. and Eakin, H. (2012). Managing private and public adaptation to climate change. *Global Environmental Change*, **22**(1), 3–11. https://doi.org/10.1016/j.gloenvcha.2011.09.010.

UNEP (2021). *Adaptation Gap Report 2021: The Gathering Storm – Adapting to Climate Change in a Post-pandemic World.* www.unep.org/resources/adaptation-gap-report-2021.

Urkidi, L. (2010). A global environmental movement against gold mining: Pascua-Lama in Chile. *Ecological Economics*, **70**(2), 219–222. https://doi.org/10.1016/j.ecolecon.2010.05.004.

Weiss, T. G. and Wilkinson, R. (2019). *Rethinking Global Governance.* Cambridge: Polity Press.

World Economic Forum (2022). *Global Risk Report 2022.* www.weforum.org/reports/global-risks-report-2022/.

WRI (2019). Aqueduct Water Risk Atlas. www.wri.org/aqueduct.

Acknowledgement

This work was supported by the Swedish Research Council for Sustainable Development (Formas) (Dnr 2022-01852). We thank Isabella Strindevall, Evelina Jonsson, and Alejandra Leon Lavandera for valuable research assistance. We are grateful to all those who contributed to this research, and in particular to the interviewees and to the participants in a manuscript workshop held in Stockholm in June 2023. We also warmly thank Robbert Biesbroek, Nina Hall, Sirkku Juhola, Andrea Nightingale, Scott Odell, and Hayley Stevenson for their valuable feedback during that workshop. Special thanks go to the series editors, Matthew Potoski and Aseem Prakash, for their invaluable support throughout the publication process. Finally, we thank the two anonymous reviewers for their constructive comments.

Cambridge Elements =

Organizational Response to Climate Change

Aseem Prakash
University of Washington

Aseem Prakash is Professor of Political Science, the Walker Family Professor for the College of Arts and Sciences, and the Founding Director of the Center for Environmental Politics at University of Washington, Seattle. His recent awards include the American Political Science Association's 2020 Elinor Ostrom Career Achievement Award in recognition of "lifetime contribution to the study of science, technology, and environmental politics," the International Studies Association's 2019 Distinguished International Political Economy Scholar Award that recognizes "outstanding senior scholars whose influence and path-breaking intellectual work will continue to impact the field for years to come," and the European Consortium for Political Research Standing Group on Regulatory Governance's 2018 Regulatory Studies Development Award that recognizes a senior scholar who has made notable "contributions to the field of regulatory governance."

Jennifer Hadden
University of Maryland

Jennifer Hadden is Associate Professor in the Department of Government and Politics at the University of Maryland. She conducts research in international relations, environmental politics, network analysis, non-state actors and social movements. Her research has been published in various journals, including the *British Journal of Political Science, International Studies Quarterly, Global Environmental Politics, Environmental Politics,* and *Mobilization.* Dr. Hadden's award-winning book, *Networks in Contention: The Divisive Politics of Global Climate Change,* was published by Cambridge University Press in 2015. Her research has been supported by a Fulbright Fellowship, as well as grants from the National Science Foundation, the National Socio-Environmental Synthesis Center, and others. She held an International Affairs Fellowship from the Council on Foreign Relations for the 2015–2016 academic year, supporting work on the Paris Climate Conference in the Office of the Special Envoy for Climate Change at the U.S. Department of State.

David Konisky
Indiana University

David Konisky is Professor at the Paul H. O'Neill School of Public and Environmental Affairs, Indiana University, Bloomington. His research focuses on U.S. environmental and energy policy, with particular emphasis on regulation, federalism and state politics, public opinion, and environmental justice. His research has been published in various journals, including *the American Journal of Political Science, Climatic Change, the Journal of Politics, Nature Energy,* and *Public Opinion Quarterly.* He has authored or edited six books on environmental politics and policy, including *Fifty Years at the U.S. Environmental Protection Agency: Progress, Retrenchment and Opportunities* (Rowman & Littlefield, 2020, with Jim Barnes and John D. Graham), *Failed Promises: Evaluating the Federal Government's Response to Environmental Justice* (MIT Press, 2015), and *Cheap and Clean: How Americans Think about Energy in the Age of Global Warming* (MIT Press, 2014, with Steve Ansolabehere). Konisky's research has been funded by the National Science Foundation, the Russell Sage Foundation, and the Alfred P. Sloan Foundation. Konisky is currently co-editor of *Environmental Politics.*

Matthew Potoski
UC Santa Barbara

Matthew Potoski is a Professor at UCSB's Bren School of Environmental Science and Management. He currently teaches courses on corporate environmental management, and his research focuses on management, voluntary environmental programs, and public policy. His research has appeared in business journals such as *Strategic Management Journal, Business Strategy and the Environment, and the Journal of Cleaner Production,* as well as public policy and management journals such as *Public Administration Review* and *the Journal of Policy Analysis and Management.* He co-authored *The Voluntary Environmentalists* (Cambridge, 2006) and *Complex Contracting* (Cambridge, 2014; the winner of the 2014 Best Book Award, American Society for Public Administration, Section on Public Administration Research) and was co-editor of *Voluntary Programs* (MIT, 2009). Professor Potoski is currently co-editor of the *Journal of Policy Analysis and Management* and the *International Public Management Journal.*

About the Series

How are governments, businesses, and non-profits responding to the climate challenge in terms of what they do, how they function, and how they govern themselves? This series seeks to understand why and how they make these choices and with what consequence for the organization and the eco-system within which it functions.

Cambridge Elements

Organizational Response to Climate Change

Elements in the Series

Explaining Transformative Change in ASEAN and EU Climate Policy: Multilevel Problems, Policies and Politics
Charanpal Bal, David Coen, Julia Kreienkamp, Paramitaningrum and Tom Pegram

Fighting Climate Change through Shaming
Sharon Yadin

Governing Sea Level Rise in a Polycentric System: Easier Said than Done
Francesca Pia Vantaggiato and Mark Lubell

Inside the IPCC: How Assessment Practices Shape Climate Knowledge
Jessica Leigh O'Reilly, Mark Vardy, Kari De Pryck and Marcela da S. Feital Benedetti

Climate Activism, Digital Technologies, and Organizational Change
Mette Eilstrup-Sangiovanni and Nina Hall

Who Tells Your Story? Women and Indigenous Peoples Advocacy at the UNFCCC
Takumi Shibaike and Bi Zhao

Climate Adaptation and Conflict Mitigation: The Case of South Sudan
Ore Koren and Jerry Urtuzuastigui

Rebel Governance in the Age of Climate Change
Kathleen Gallagher Cunningham, Leonardo Gentil-Fernandes, Elisabeth Gilmore, Reyko Huang, Danielle F. Jung and Cyanne E. Loyle

Greening the International Monetary Fund
Alexandros Kentikelenis and Thomas Stubbs

Corporate Climate Adaptation: Translating Complex Societal Risks into Business as Usual
Maria-Therese Gustafsson and Lisa Dellmuth

A full series listing is available at: www.cambridge.org/ORCC

For EU product safety concerns, contact us at Calle de José Abascal, 56–1°,
28003 Madrid, Spain or eugpsr@cambridge.org.

www.ingramcontent.com/pod-product-compliance
Ingram Content Group UK Ltd.
Pitfield, Milton Keynes, MK11 3LW, UK
UKHW022300240426
470365UK00008B/172